IMS-DB Basic Training

For Application Developers

Robert Wingate

ISBN: 9781734584752

Disclaimer

The contents of this book are based upon the author's understanding of and experience with the IBM IMS-DB product. Every attempt has been made to provide correct information. However, the author and publisher do not guarantee the accuracy of every detail, nor do they assume responsibility for information included in or omitted from it. All of the information in this book should be used at your own risk.

Copyright

The contents of this book may not be copied in whole, or in part, without the explicit written permission of the author. The contents are intended for personal use only. Secondary distribution for gain is not allowed. Any alteration of the contents is absolutely forbidden.

Table of Contents

Introduction

Welcome

Congratulations on your purchase of **IMS Basic Training for Application Developers.** This book will teach you the basic information and skills you need to develop IMS applications with COBOL and PLI on IBM mainframes running z/OS. The instruction, examples and sample programs in this book are a fast track to becoming productive as quickly as possible using IMS with COBOL and PLI. The content is easy to read and digest, well organized and focused on honing real job skills.

This is not an "everything you need to know about IMS" book. Rather, this text will teach you what you need to know to become **productive quickly** with IMS. For additional detail, you can download and reference the IBM manuals and Redbooks associated with these products.

Assumptions:

While I do not assume that you know a great deal about IBM mainframe programming, I do assume that you've logged into an IBM mainframe and know your way around. Also I assume that you have a working knowledge of computer programming in either COBOL or PLI. All in all, I assume you have:

1. A working knowledge of ISPF navigation and basic operations such as creating data sets.

2. A rudimentary understanding of hierarchical database concepts.

3. A working understanding of the COBOL or PLI programming language.

4. Access to a mainframe computer running z/OS and DB2 (with a COBOL compiler available).

Approach to Learning

I suggest you follow along and do the examples yourself in your own test environment. There's nothing like hands-on experience. Going through the motions will help you learn faster.

If you do not have access to a mainframe system through your job, I can recommend Mathru Technologies. You can rent a mainframe account from them at a very affordable rate, and this includes access to IMS if you request it. Their environment supports COBOL and PLI as well. The URL to the Mathru web site is:

http://mathrutech.com/index.html

Besides the instruction and examples, I've included questions at the end of each chapter. I recommend that you answer these and then check yourself against the answers in the back of the book.

Knowledge, experience and practice questions. Will that guarantee that you'll succeed as an IMS application developer? Of course, nothing is guaranteed in life. But if you put sufficient effort into this well-rounded training plan that includes all three of the above, I believe you have a very good chance of becoming productive as an IMS Application Developer as soon as possible.

Best of luck!

Robert Wingate
IBM Certified Application Developer – DB2 11 for z/OS

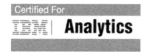

IMS Background

IMS is a hierarchical database management system (DBMS) that has been around since the 1960's. Although relational DBMSs are more common now, there is still an installed base of IMS users and IBM provides robust support for it. IMS is highly tuned for transaction management and generally provides excellent performance for that environment.

This text deals with IMS-DB, the IMS database manager. IMS also has a transaction manager called IMS-DC. We will be covering IMS-DB in this volume, and IMS-DC in later volume.

There are two modes of running IMS programs. One is DLI which runs within its own address space. There is also Batch Mode Processing (BMP) which runs under the IMS online control region. The practical difference between the two concerns programs that update the database. In DLI mode, a program requires exclusive use of the database to perform updates. In BMP mode, a program does not require exclusive use of the database because it is run in the shared IMS online environment. The IMS online system "referees" the shared online environment.

Before going further I need to point out that in IMS data records are called "segments". I'll use the terms segment and record more or less interchangeably throughout the chapter. There are usually multiple segment types in an IMS database, although not always.

With our common understanding of these basic terms, let's move forward and design/ build a simple IMS database.

Designing and Creating IMS Databases

Sample System Specification

We're going to create a hierarchical database for a Human Resource system that will involve employees. In fact the database will be named EMPLOYEE and the root segment (highest level segment type) will also be named EMPLOYEE. This segment will include information such as name, years of service and last promotion date.

The EMPLOYEE segment will have a child segment that stores details about the employee's pay. The segment will be named EMPPAY and include the effective date, annual pay and bonus pay.

The EMPPAY segment will have a child segment named EMPPAYHS that includes historical details about each paycheck an employee received.

Note: there can be multiple EMPPAY segments under each EMPLOYEE segment, and there can be multiple EMPPAYHS segments under each EMPPAY segment. The following diagram depicts our EMPLOYEE database visually as a hierarchy.

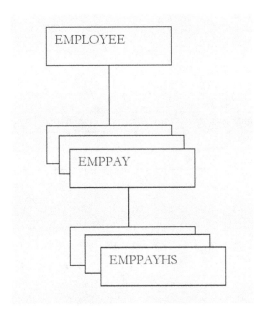

The following shows the segment structure we will be using to organize the three record types. Note that the EMP_ID key is only required on the root segment. You cannot access the child segments except through the root, so this makes sense.

EMPLOYEE Segment (key is EMP_ID).

Field Name	Type
EMP_ID	**INTEGER**
EMP_LAST_NAME	VARCHAR(30)
EMP_FIRST_NAME	VARCHAR(20)
EMP_SERVICE_YEARS	INTEGER
EMP_PROMOTION_DATE	DATE

EMPPAY segment (key is EFF_DATE which means effective date):

Field Name	Type
EFF_DATE	**DATE**
EMP_REGULAR_PAY	DECIMAL
EMP_BONUS_PAY	DECIMAL
EMP_SEMIMTH_PAY	DECIMAL

EMPPAYHS segment (key is PAY_DATE):

Field Name	Type
PAY_DATE	**DATE**
ANNUAL_PAY	DECIMAL
PAY_CHECK_AMT	DECIMAL

Having decided the content of our segment types, we can now create a record layout for each of these. We'll do this in COBOL since we will be writing our IMS programs in that language. We'll also create the corresponding PLI record layouts when we write the PLI programs.

Here's the layout for the EMPLOYEE segment:

```
01 IO-EMPLOYEE-RECORD.
    05  EMP-ID          PIC X(04).
    05  FILLER          PIC X(01).
    05  EMPL-LNAME      PIC X(30).
    05  FILLER          PIC X(01).
    05  EMPL-FNAME      PIC X(20).
    05  FILLER          PIC X(01).
    05  EMPL-YRS-SRV    PIC X(02).
    05  FILLER          PIC X(01).
    05  EMPL-PRM-DTE    PIC X(10).
    05  FILLER          PIC X(10).
```

We've provided a bit of filler between fields, and we've left 10 bytes at the end (yes later we will be adding a field so we need some free space). Our total is 80 bytes for this segment type. The record will be keyed on **EMP-ID** which is the first four bytes of the record. We'll need this information to define the database. Next, the EMPPAY segment layout is as follows:

```
01 IO-EMPPAY-RECORD.
    05  PAY-EFF-DATE  PIC X(8).
    05  PAY-REG-PAY   PIC S9(6)V9(2) USAGE COMP-3.
    05  PAY-BON-PAY   PIC S9(6)V9(2) USAGE COMP-3.
    05  SEMIMTH-PAY   PIC S9(6)V9(2) USAGE COMP-3.
    05  FILLER        PIC X(57).
```

Notice there is no EMP-ID field. As mentioned earlier, child segments do not need to repeat the parent segment key. The hierarchical structure of the database makes this unnecessary. The PAY-EFF-DATE field will be the key for the EMPPAY segment, and it is 8 bytes. The format will be YYYYMMDD.

Also notice that we padded the record with filler to total 80 bytes. We didn't have to do this. The record size is actually 23 bytes without the filler. But often it's a convenience to leave space in the IO layout for future expansion.

Finally, here is the layout for the EMPPAYHS segment. Out key will be PAY-DATE and it will be formatted as YYYYMMDD.

```
01 IO-EMPPAYHS-RECORD.
    05  PAY-DATE      PIC X(8).
    05  PAY-ANN-PAY   PIC S9(6)V9(2) USAGE COMP-3.
    05  PAY-AMT       PIC S9(6)V9(2) USAGE COMP-3.
    05  FILLER        PIC X(57).
```

Now we are ready to build the data base descriptor!

Database Descriptor (DBD)

A database descriptor is required to have an IMS database. The descriptor specifies the name of the database, plus the various segment types. Typically a database administrator will create and maintain DBDs. You should still understand how to read the DBD code to understand the structure of the database.

Here's the DBD code for our EMPLOYEE database.

```
PRINT NOGEN
DBD NAME=EMPLOYEE,ACCESS=HISAM
DATASET DD1=EMPLOYEE,OVFLW=EMPLFLW
SEGM NAME=EMPLOYEE,PARENT=0,BYTES=80
FIELD NAME=(EMPID,SEQ,U),BYTES=04,START=1,TYPE=C
SEGM NAME=EMPPAY,PARENT=EMPLOYEE,BYTES=23
FIELD NAME=(EFFDATE,SEQ,U),START=1,BYTES=8,TYPE=C
SEGM  NAME=EMPPAYHS,PARENT=EMPPAY,BYTES=18
FIELD NAME=(PAYDATE,SEQ,U),START=1,BYTES=8,TYPE=C
DBDGEN
FINISH
END
```

The code above specifies the name of the database which is EMPLOYEE, as well as an access method of HISAM (Hierarchical Indexed Sequential Access Method). HISAM database records are stored in two data sets: a primary data set and an overflow data set. The primary dataset is always a VSAM KSDS and the overflow dataset is a VSAM ESDS. The ESDS dataset is used if the KSDS dataset becomes full. In that case any new records are inserted to the (overflow) ESDS dataset.

There is considerable information available on the IBM web site about how HISAM records are stored, but that information is frankly not very useful for application programmer duties. If you are interested, you can find it here. [1]

Looking at the DBD code, we see that the DATASET DD1 and OVFLW keywords define the DD names of the primary cluster and the overflow dataset, respectively. We defined these values as EMPLOYEE and EMPLFLW. Later when we run batch jobs against the database, the DD name in our JCL must be EMPLOYEE for the KSDS file, and EMPLFLW

---

1 https://www.ibm.com/support/knowledgecenter/en/SSEPH2_13.1.0/com.ibm.ims13.doc.dag/ims_hisamdb.htm

for the overflow dataset.

Next, we define our segment types using the `SEGM NAME=` keywords. We also define the parent of each segment unless the segment is the root segment in which case we specify `PARENT=0`. Next you specify the length of your segment. We've defined the length as the total of the fields we mapped out earlier in the COBOL layouts.

For each segment, if you have any searchable fields (such as keys), they must be defined with the `FIELD NAME=` keywords. In our case, we will only specify the key fields for each segment. W specify the field name, that the records are to be ordered sequentially (SEQ), and that the field content must be unique(U). Then we specify how many bytes the field is, and it's displacement in the record. We also specify C for character data – the actual data we store can be of any type but we specify C to indicate the default type is character data.

Here's an example of defining the employee id in the DBD from above:

```
FIELD NAME=(EMPID,SEQ,U),BYTES=04,START=1,TYPE=C
```

Finally, you conclude the DBD with

```
DBDGEN
FINISH
END
```

Now you are ready to run your installation's JCL to generate a `DBD`. Most likely you will ask a DBA to do this. Here's the JCL I run which executes a proc named `DBDGEN`. It will be different for your installation.

```
//USER01D JOB MSGLEVEL=(1,1),NOTIFY=&SYSUID
//*
//PLIB    JCLLIB ORDER=SYS1.IMS.PROCLIB
//DGEN    EXEC DBDGEN,
//            MEMBER=EMPLOYEE,              <= DBD SOURCE MEMBER
//            SRCLIB=USER01.IMS.SRCLIB,     <= DBD SOURCE LIBRARY
//            DBDLIB=USER01.IMS.DBDLIB      <= DBD LIBRARY
//*
```

More information about designing, coding and generating DBDs is available on the

IBM product web site.

Supporting VSAM Files

Now that we have a DBD generated, we can create the physical files for the database. Actually it can be done in either order, but you do need to know the maximum record size for all segments in order to build the IDCAMS JCL. For IMS datasets we use a VSAM key sequenced data set (KSDS).

Here is the JCL for creating our EMPLOYEE IMS database. Notice that we specify a RECORDSIZE that is 8 bytes longer than the logical record size that we defined in the DBD. And although the key is the first logical byte of each record, we specify the key displacement at byte 6. The IMS system uses the first 5 bytes of each record, so this is required (IMS also uses the last 3 bytes, so we end up with 8 additional bytes for the record size).

Finally, note that we have a second job step to repro a dummy file to our VSAM cluster name to initialize it. This is required. Otherwise we will get an abend the first time we try to access it. Go ahead and run the JCL or ask your DBA to create the physical files.

```
//USER01D JOB MSGLEVEL=(1,1),NOTIFY=&SYSUID
//*
//***************************************************************
//* DEFINE VSAM KSDS CLUSTER FOR EMPLOYEE DATABASE
//***************************************************************
//VDEF     EXEC PGM=IDCAMS
//SYSPRINT DD SYSOUT=*
//SYSIN    DD  *
   DEFINE CLUSTER(NAME(USER01.IMS.EMPLOYEE.CLUSTER)   -
                 INDEXED                              -
                 KEYS(4,6)                            -
                 RECORDSIZE(88,88)                    -
                 TRACKS(2,1)                          -
                 CISZ(2048)                           -
                 VOLUMES(DEVHD1)                      -
                 )                                    -
          DATA(NAME(USER01.IMS.EMPLOYEE.DATA))
//*
//***************************************************************
//* INITIALIZE THE VSAM FILE TO PLACE EOF MARK
//***************************************************************
//VINIT    EXEC PGM=IDCAMS
```

```
//SYSPRINT DD  SYSOUT=*
//INF       DD  DUMMY
//OUTF      DD  DSN=USER01.IMS.EMPLOYEE.CLUSTER,DISP=SHR
//SYSIN     DD  *
  REPRO INFILE(INF) OUTFILE(OUTF)
/*
//*
```

Next, here is the JCL for creating the overflow dataset. When your KSDS is full, new records will be placed in the overflow dataset. Again, specify RECORDSIZE 88 (not 80). You must also specify NONINDEXED to get an ESDS file.

```
//USER01D JOB MSGLEVEL=(1,1),NOTIFY=&SYSUID
//*
//************************************************************
//* DEFINE VSAM ESDS CLUSTER FOR IMS DATA BASE
//************************************************************
//VDEF      EXEC PGM=IDCAMS
//SYSPRINT DD  SYSOUT=*
//SYSOUT   DD  SYSOUT=*
//SYSIN     DD  *
  DEFINE CLUSTER(NAME(USER01.IMS.EMPLFLW.CLUSTER)  -
                 NONINDEXED                         -
                 RECORDSIZE(88,88)                  -
                 TRACKS(2,1)                        -
                 CISZ(2048)                         -
                 VOLUMES(DEVHD1)                    -
                 )                                  -
        DATA(NAME(USER01.IMS.EMPLFLW.DATA))
/*
```

Ok, time to move on to our next IMS entity which is a PSB.

Program Specification Block (PSB)

A Program Specification Block (PSB) is an IMS entity that specifies which segments and operations can be performed on one or more databases (using this particular PSB authority). PSBs consist of one or more Program Communication Blocks (PCB) which are logical views of a database. It is typical for each IMS application program to have a separate PSB defined for it, but this is convention, not an IMS requirement. For our programming examples we will mostly use just one PSB, but we will modify it a few times. Here is the code for the PSB that we will be using for most of the examples.

```
PRINT NOGEN
PCB    TYPE=DB,NAME=EMPLOYEE,KEYLEN=20,PROCOPT=AP
SENSEG NAME=EMPLOYEE,PARENT=0
SENSEG NAME=EMPPAY,PARENT=EMPLOYEE
SENSEG NAME=EMPPAYHS,PARENT=EMPPAY
PSBGEN LANG=COBOL,PSBNAME=EMPLOYEE
END
```

Here's the meaning of each keyword for defining the PCB.

PCB – this is where you define a pointer to your database.

TYPE - typically this is DB to indicate a database PCB which provides access to a specific database. There is also a terminal (TP) PCB that is used for teleprocessing calls in IMS DC, but we won't be doing IMS DC in this text.

NAME - identifies the database to be accessed.

PROCOPT - Processing options. This value specifies which operations can be performed such as read, update or delete. The following are the most common options:

G   Get
I   Insert
R   Replace
D   Delete
A   All Options (G, I, R, D)
L   Load Function (Initial Loading)
LS   Load Function (Loading Sequentially)
K   Key Function - Access only key of the segment.
O   Used with G option to indicate that HOLD is not allowed.
P   Path Function (Used during Path Calls)

The PROCOPT can be defined for the entire PCB or it can be more granular by applying it to specific segments. If specified at the segment level, it overrides any PROCOPT at the PCB level. In our case we have specified PROCOPT=AP for the entire PSB. That is powerful. It means All (G, I, R, D) plus authority to do path calls.

KEYLEN – specifies the length of the concatenated key. Concatenated key is the maximum length of all the segment keys added up. This needs to be calculated by adding the longest segment key in each level from top to bottom.

SENSEG means sensitive segment, which means you can access that segment via this PSB. You can specify which segments you want to access. You might not always want all segments to be accessed. In our case we do, so we define a SENSEG for each segment type.

You must then execute a PSBGEN (or ask your DBA to). Here is the JCL I use which executes a proc named PSBGEN, and note that the member name I stored the PSB source under is EMPPSB. Your JCL will be different and specific to the installation:

```
//USER01D JOB MSGLEVEL=(1,1),NOTIFY=&SYSUID
//*
//PLIB    JCLLIB ORDER=SYS1.IMS.PROCLIB
//PGEN    EXEC PSBGEN,
//             MEMBER=EMPPSB,             <= PSB SOURCE MEMBER
//             SRCLIB=USER01.IMS.SRCLIB,  <= PSB SOURCE LIBRARY
//             PSBLIB=USER01.IMS.PSBLIB   <= PSB LIBRARY
//*
```

Now you have the basic building blocks of IMS – a database descriptor (DBD), the physical VSAM files to support the database, and a PSB that provides permissions to access the data within the database via PCBs. We are ready to start programming.

IMS Application Programming Basics

The IMS Program Interface

To request IMS data services in an application program, you must call the IMS interface program for that programming language. The interface program for COBOL is CBLTDLI. This program is called with several parameters which vary depending on the operation being requested. The call also needs to tell CBLTDLI how many parameters are being passed, so we'll declare some constants in our program for that.

```
01 IMS-RET-CODES.
    05 THREE          PIC S9(9) COMP VALUE +3.
    05 FOUR           PIC S9(9) COMP VALUE +4.
    05 FIVE           PIC S9(9) COMP VALUE +5.
    05 SIX            PIC S9(9) COMP VALUE +6.
```

Let's take an example where you want to retrieve an EMPLOYEE segment from the EMPLOYEE database, and you want employee number 3217. Here is the call with appropriate parameters. We'll discuss each of these in turn.

```
CALL 'CBLTDLI' USING FOUR,
               DLI-FUNCGU,
               PCB-MASK,
               SEG-IO-AREA,
               EMP-QUALIFIED-SSA
```

The first parameter specifies the number of parameters being passed. In the case of the above call, the number would be four.

The second parameter is the call type. The following are the common IMS calls used to insert, retrieve, modify and delete data in an IMS database. There are some other calls we'll introduce later such as for checkpointing and rolling back data changes.

CALL	CALL DESCRIPTION
DLET	The Delete (DLET) call is used to remove a segment and its dependents from the database.

GN/GHN	The Get Next (GN) call is used to retrieve segments sequentially from the database. The Get Hold Next (GHN) is the hold form for a GN call.
GNP/GHNP	The Get Next in Parent (GNP) call retrieves dependents sequentially. The Get Hold Next in Parent (GHNP) call is the hold form of the GNP call.
GU/GHU	The Get Unique (GU) call is used to directly retrieve segments and to establish a starting position in the database for sequential processing. The Get Hold Unique (GHU) is the hold form for a GU call.
ISRT	The Insert (ISRT) call is used to load a database and to add one or more segments to the database. You can use ISRT to add a record to the end of a GSAM database or for an alternate PCB that is set up for IAFP processing.
REPL	The Replace (REPL) call is used to change the values of one or more fields in a segment.

It is a common practice to define a set of constants in your program that specify the value of the specific IMS calls. Here's the COBOL code to put in working storage for this purpose.

```
01  DLI-FUNCTIONS.
    05  DLI-FUNCISRT   PIC X(4)  VALUE  'ISRT'.
    05  DLI-FUNCGU     PIC X(4)  VALUE  'GU  '.
    05  DLI-FUNCGN     PIC X(4)  VALUE  'GN  '.
    05  DLI-FUNCGHU    PIC X(4)  VALUE  'GHU '.
    05  DLI-FUNCGHN    PIC X(4)  VALUE  'GHN '.
    05  DLI-FUNCGNP    PIC X(4)  VALUE  'GNP '.
    05  DLI-FUNCREPL   PIC X(4)  VALUE  'REPL'.
    05  DLI-FUNCDLET   PIC X(4)  VALUE  'DLET'.
    05  DLI-FUNCXRST   PIC X(4)  VALUE  'XRST'.
    05  DLI-FUNCCHKP   PIC X(4)  VALUE  'CHKP'.
    05  DLI-FUNCROLL   PIC X(4)  VALUE  'ROLL'.
```

As you can see from the call above and the constant definitions, we are doing a Get Unique (GU) call. The `DLI-FUNCGU` specifies it.

The next parameter is a `PCB` data area that we defined as `PCB-MASK`. This returns various information from IMS after the database call. You must define this structure in the Linkage Section of your program since it is passing data back and forth from the `CBLTDLI` interface program.

```
LINKAGE SECTION.
01  PCB-MASK.
    03  DBD-NAME        PIC X(8).
    03  SEG-LEVEL       PIC XX.
    03  STATUS-CODE     PIC XX.
    03  PROC-OPT        PIC X(4).
    03  FILLER          PIC X(4).
    03  SEG-NAME        PIC X(8).
    03  KEY-FDBK        PIC S9(5)  COMP.
    03  NUM-SENSEG      PIC S9(5)  COMP.
    03  KEY-FDBK-AREA.
        05  EMPLOYEE-KEY  PIC X(04).
```

One of the most important data elements in the `PCB-MASK` is the two byte status code returned by the call, the `STATUS-CODE`. A blank status code means that the call was successful. Other status codes indicate the reason why the call failed. Here is a subset of the status codes you may encounter. The complete list of codes is found on the IBM web site. [2]

---

2 https://www.ibm.com/support/knowledgecenter/en/SSEPH2_13.1.0/com.ibm.ims13.doc.msgs/compcodes/ims_dlistatuscodestables.htm#ims_dlistatuscodestables__catdlistatuscodes

IMS Status Codes

PCB Status Code	Description
AC	Hierarchic error in SSAs.
AD	Function parameter incorrect. Only applies to full-function DEQ calls.
AI	Data management OPEN error.
AJ	Incorrect parameter format in I/O area; incorrect SSA format; incorrect command used to insert a logical child segment. I/O area length in AIB is invalid; incorrect class parameter specified in Fast Path Q command code.
AK	Invalid SSA field name.
AM	Call function not compatible with processing option, segment sensitivity, transaction code, definition, or program type.
AU	SSAs too long.
DA	Segment key field or non-replaceable field has been changed.
DJ	No preceding successful GHU or GHN call or an SSA supplied at a level not retrieved.
FT	Too many SSAs on call.
GB	End of database.
GE	Segment not found.
GG	Segment contains invalid pointer.
GP	No parentage established.
II	Segment already exists.

In your application program control is passed from IMS through an entry point. Your entry point must refer to the PCBs in the order in which they have been defined in the PSB. When you code each DL/I call, you must provide the PCB you want to use for that call. Here is the entry point code at the beginning of the procedure division for this program.

```
ENTRY 'DLITCBL' USING PCB-MASK
```

The next parameter is the segment I/O area. This is where IMS returns the data segment you requested, or where you load data to be inserted/updated on an insert or replace command. For the `EMPLOYEE` record, we will define the I/O area in COBOL as:

```
01 IO-EMPLOYEE-RECORD.
   05  EMP-ID         PIC X(04).
   05  FILLER         PIC X(01).
   05  EMPL-LNAME     PIC X(30).
   05  FILLER         PIC X(01).
   05  EMPL-FNAME     PIC X(20).
   05  FILLER         PIC X(01).
   05  EMPL-YRS-SRV   PIC X(02).
   05  FILLER         PIC X(01).
   05  EMPL-PRM-DTE   PIC X(10).
   05  FILLER         PIC X(10).
```

The last parameter in our call is a Segment Search Argument (SSA). This is where we specify the type of segment we want and the key value. It is also possible to simply request the next record of a particular segment type without regard to key value. When we specify a key, that means we are using a "qualified" SSA. When we don't specify a key, it means we are using an unqualified SSA.

Here's the COBOL definition of the qualified and unqualified SSAs for the `EMPLOYEE` segment.

```
01 EMP-QUALIFIED-SSA.
   05  SEGNAME      PIC X(08) VALUE 'EMPLOYEE'.
   05  FILLER       PIC X(01) VALUE '('.
   05  FIELD        PIC X(08) VALUE 'EMPID'.
   05  OPER         PIC X(02) VALUE ' ='.
   05  EMP-ID-VAL   PIC X(04) VALUE '    '.
   05  FILLER       PIC X(01) VALUE ')'.

01 EMP-UNQUALIFIED-SSA.
   05  SEGNAME      PIC X(08) VALUE 'EMPLOYEE'.
   05  FILLER       PIC X(01) VALUE ' '.
```

Both qualified and unqualified SSAs must specify the segment type or name. You specify the key for the qualified SSA in the field we've named `EMP-ID-VAL`. We'll show many examples of SSAs in the program examples, including the use of Boolean SSA values.

Loading the IMS Database

Ok, finally to our first program. We're going to load the IMS database with a few records from a text file (a.k.a. a flat file). Here is the data file contents:

```
----+----1----+----2----+----3----+----4----+----5----+----6----+----7----+----8
******************************* Top of Data ********************************
1111 VEREEN             CHARLES             12 2017-01-01 937253058
1122 JENKINS            DEBORAH             05 2017-01-01 435092366
3217 JOHNSON            EDWARD              04 2017-01-01 397342007
4175 TURNBULL           FRED                01 2016-12-01 542083017
4720 SCHULTZ            TIM                 09 2017-01-01 650450254
4836 SMITH              SANDRA              03 2017-01-01 028374669
6288 WILLARD            JOE                 06 2016-01-01 209883920
7459 STEWART            BETTY               07 2016-07-31 019572830
9134 FRANKLIN           BRIANNA             00 2016-10-01 937293598
```

As you can see, we've formatted the records exactly like we want them to be applied to the database. Of course, your input layout could be different than the IMS segment layout, but using the same layout makes it easier because you don't have to do field assignments in the program.

Now let's create a program named COBIMS1 to load the data. We'll define the input file of course. In our program, let's call the file EMPFILE. Let's assume that the DD name for the employee load file is EMPIFILE.

```
ENVIRONMENT DIVISION.
INPUT-OUTPUT SECTION.
FILE-CONTROL.
    SELECT EMPFILE ASSIGN TO EMPIFILE.

DATA DIVISION.
FILE SECTION.
FD  EMPFILE
    RECORDING MODE IS F
    RECORD CONTAINS 80 CHARACTERS.

01 INSRT-REC.
    05 SEG-IO-AREA PIC X(80).
```

We've specified a SEG-IO-AREA variable to read the input file into and to write the IMS record from. We could have used the fully detailed IO-EMPLOYEE-RECORD instead (and we will later), but I want to demonstrate the value of having your input records structured the same as the IMS segment. When you do this, it really simplifies the coding such that you can both read and write using use a one element structure like `SEG-IO-AREA`.

Next we'll code the working storage section with a few things including:

1. An end of file switch for the loop we'll create to load the records.

2. The DLI call constants.

3. The Employee segment I/O structure.

4. The Employee segment SSA.

Here it is:

```
WORKING-STORAGE SECTION.

  01 WS-FLAGS.
      05  SW-END-OF-FILE-SWITCH    PIC X(1) VALUE 'N'.
          88  SW-END-OF-FILE                VALUE 'Y'.
          88  SW-NOT-END-OF-FILE            VALUE 'N'.

  01 DLI-FUNCTIONS.
      05 DLI-FUNCISRT  PIC X(4) VALUE 'ISRT'.
      05 DLI-FUNCGU    PIC X(4) VALUE 'GU  '.
      05 DLI-FUNCGN    PIC X(4) VALUE 'GN  '.
      05 DLI-FUNCGHU   PIC X(4) VALUE 'GHU '.
      05 DLI-FUNCGNP   PIC X(4) VALUE 'GNP '.
      05 DLI-FUNCREPL  PIC X(4) VALUE 'REPL'.
      05 DLI-FUNCDLET  PIC X(4) VALUE 'DLET'.
      05 DLI-FUNCXRST  PIC X(4) VALUE 'XRST'.
      05 DLI-FUNCCKPT  PIC X(4) VALUE 'CKPT'.

  01 IO-EMPLOYEE-RECORD.
      05  EMPL-ID-IN    PIC X(04).
      05  FILLER        PIC X(01).
      05  EMPL-LNAME    PIC X(30).
      05  FILLER        PIC X(01).
      05  EMPL-FNAME    PIC X(20).
      05  FILLER        PIC X(01).
      05  EMPL-YRS-SRV  PIC X(02).
      05  FILLER        PIC X(01).
      05  EMPL-PRM-DTE  PIC X(10).
      05  FILLER        PIC X(10).
```

```
01  EMP-UNQUALIFIED-SSA.
    05  SEGNAME     PIC X(08) VALUE 'EMPLOYEE'.
    05  FILLER      PIC X(01) VALUE ' '.

01  EMP-QUALIFIED-SSA.
    05  SEGNAME     PIC X(08) VALUE 'EMPLOYEE'.
    05  FILLER      PIC X(01) VALUE '('.
    05  FIELD       PIC X(08) VALUE 'EMPID'.
    05  OPER        PIC X(02) VALUE ' ='.
    05  EMP-ID-VAL  PIC X(04) VALUE '    '.
    05  FILLER      PIC X(01) VALUE ')'.

01  IMS-RET-CODES.
    05  THREE       PIC S9(9) COMP VALUE +3.
    05  FOUR        PIC S9(9) COMP VALUE +4.
    05  FIVE        PIC S9(9) COMP VALUE +5.
    05  SIX         PIC S9(9) COMP VALUE +6.
```

Finally, we'll code the linkage section which includes the database PCB mask.

```
LINKAGE SECTION.
01 PCB-MASK.
    03 DBD-NAME        PIC X(8).
    03 SEG-LEVEL       PIC XX.
    03 STATUS-CODE     PIC XX.
    03 PROC-OPT        PIC X(4).
    03 FILLER          PIC X(4).
    03 SEG-NAME        PIC X(8).
    03 KEY-FDBK        PIC S9(5) COMP.
    03 NUM-SENSEG      PIC S9(5) COMP.
    03 KEY-FDBK-AREA.
       05 EMPLOYEE-KEY  PIC X(04).
```

We'll work on the procedure division next, which will complete the program. Let's talk about the actual database call. Here's what we'll use:

```
CALL 'CBLTDLI' USING FOUR,
     DLI-FUNCISRT,
     PCB-MASK,
     SEG-IO-AREA,
     EMP-UNQUALIFIED-SSA
```

This is similar to the example we gave earlier with a couple of differences. One difference of course is that we are doing an `ISRT` call, so we specify the constant `DLI-FUN-CISRT`. The other difference is that we will use an unqualified SSA. On an insert operation, IMS will always establish the record key from the I/O area and therefore it does not use a qualified SSA.

To be clear, any time you are inserting a record, you will use an **unqualified** SSA at the level of the record you are inserting. So if you are inserting a root segment, you will always use an unqualified SSA. If you are inserting a child segment under a root, you will use a qualified SSA on the root segment, and then an unqualified SSA for the child segment. If this seems a bit cryptic now, it should make more sense in later examples where we use child segments and multiple SSAs.

Ok, here's our complete program code. See what you think.

```
       IDENTIFICATION DIVISION.
       PROGRAM-ID. COBIMS1.

      ******************************************************
      *   INSERT A RECORD INTO IMS EMPLOYEE DATABASE        *
      ******************************************************

       ENVIRONMENT DIVISION.
       INPUT-OUTPUT SECTION.
       FILE-CONTROL.
           SELECT EMPFILE ASSIGN TO EMPIFILE.

       DATA DIVISION.
       FILE SECTION.
       FD  EMPFILE
           RECORDING MODE IS F
           RECORD CONTAINS 80 CHARACTERS.

       01 INSRT-REC.
          05 SEG-IO-AREA PIC X(80).

      ******************************************************
      *   W O R K I N G   S T O R A G E   S E C T I O N     *
      ******************************************************

       WORKING-STORAGE SECTION.

       01 WS-FLAGS.
          05  SW-END-OF-FILE-SWITCH   PIC X(1) VALUE 'N'.
              88  SW-END-OF-FILE                VALUE 'Y'.
```

```
              88  SW-NOT-END-OF-FILE              VALUE 'N'.


    01 DLI-FUNCTIONS.
       05 DLI-FUNCISRT  PI X(4) VALUE 'ISRT'.
       05 DLI-FUNCGU    PIC X(4) VALUE 'GU  '.
       05 DLI-FUNCGN    PIC X(4) VALUE 'GN  '.
       05 DLI-FUNCGHU   PIC X(4) VALUE 'GHU '.
       05 DLI-FUNCGNP   PIC X(4) VALUE 'GNP '.
       05 DLI-FUNCREPL  PIC X(4) VALUE 'REPL'.
       05 DLI-FUNCDLET  PIC X(4) VALUE 'DLET'.
       05 DLI-FUNCXRST  PIC X(4) VALUE 'XRST'.
       05 DLI-FUNCCKPT  PIC X(4) VALUE 'CKPT'.


    01 IN-EMPLOYEE-RECORD.
       05  EMPL-ID-IN    PIC X(04).
       05  FILLER        PIC X(01).
       05  EMPL-LNAME    PIC X(30).
       05  FILLER        PIC X(01).
       05  EMPL-FNAME    PIC X(20).
       05  FILLER        PIC X(01).
       05  EMPL-YRS-SRV  PIC X(02).
       05  FILLER        PIC X(01).
       05  EMPL-PRM-DTE  PIC X(10).
       05  FILLER        PIC X(10).


   01 EMP-UNQUALIFIED-SSA.
      05  SEGNAME     PIC X(08) VALUE 'EMPLOYEE'.
      05  FILLER      PIC X(01) VALUE ' '.


   01 EMP-QUALIFIED-SSA.
      05  SEGNAME     PIC X(08) VALUE 'EMPLOYEE'.
      05  FILLER      PIC X(01) VALUE '('.
      05  FIELD       PIC X(08) VALUE 'EMPID'.
      05  OPER        PIC X(02) VALUE ' ='.
      05  EMP-ID-VAL  PIC X(04) VALUE '    '.
      05  FILLER      PIC X(01) VALUE ')'.


   01 IMS-RET-CODES.
      05 THREE        PIC S9(9) COMP VALUE +3.
      05 FOUR         PIC S9(9) COMP VALUE +4.
      05 FIVE         PIC S9(9) COMP VALUE +5.
      05 SIX          PIC S9(9) COMP VALUE +6.
```

```
LINKAGE SECTION.
 01 PCB-MASK.
     03 DBD-NAME         PIC X(8).
     03 SEG-LEVEL        PIC XX.
     03 STATUS-CODE      PIC XX.
     03 PROC-OPT         PIC X(4).
     03 FILLER           PIC X(4).
     03 SEG-NAME         PIC X(8).
     03 KEY-FDBK         PIC S9(5) COMP.
     03 NUM-SENSEG       PIC S9(5) COMP.
     03 KEY-FDBK-AREA.
        05 EMPLOYEE-KEY  PIC X(04).
        05 EMPPAYHS-KEY  PIC X(08).


 PROCEDURE DIVISION.

     INITIALIZE PCB-MASK
     ENTRY 'DLITCBL' USING PCB-MASK

     PERFORM P100-INITIALIZATION.
     PERFORM P200-MAINLINE.
     PERFORM P300-TERMINATION.
     GOBACK.


 P100-INITIALIZATION.

     DISPLAY '** PROGRAM COBIMS1 START **'
     DISPLAY 'PROCESSING IN P100-INITIALIZATION'
     OPEN INPUT EMPFILE.


 P200-MAINLINE.

     DISPLAY 'PROCESSING IN P200-MAINLINE'


     READ EMPFILE
        AT END SET SW-END-OF-FILE TO TRUE
     END-READ


     PERFORM UNTIL SW-END-OF-FILE

        CALL 'CBLTDLI' USING FOUR,
             DLI-FUNCISRT,
             PCB-MASK,
             SEG-IO-AREA,
             EMP-UNQUALIFIED-SSA
```

35

```
        IF STATUS-CODE = '  '
            DISPLAY 'SUCCESSFUL INSERT-REC:' SEG-IO-AREA
        ELSE
            PERFORM P400-DISPLAY-ERROR
        END-IF

        READ EMPFILE
            AT END SET SW-END-OF-FILE TO TRUE
        END-READ

    END-PERFORM.



    P300-TERMINATION.

        DISPLAY 'PROCESSING IN P300-TERMINATION'

        CLOSE EMPFILE
        DISPLAY '** COBIMS1 - SUCCESSFULLY ENDED **'.


    P400-DISPLAY-ERROR.

        DISPLAY 'ERROR ENCOUNTERED - DETAIL FOLLOWS'
        DISPLAY 'SEG-IO-AREA      :' SEG-IO-AREA
        DISPLAY 'DBD-NAME1:'      DBD-NAME
        DISPLAY 'SEG-LEVEL1:'     SEG-LEVEL
        DISPLAY 'STATUS-CODE:'    STATUS-CODE
        DISPLAY 'PROC-OPT1 :'     PROC-OPT
        DISPLAY 'SEG-NAME1 :'     SEG-NAME
        DISPLAY 'KEY-FDBK1 :'     KEY-FDBK
        DISPLAY 'NUM-SENSEG1:'    NUM-SENSEG
        DISPLAY 'KEY-FDBK-AREA1:' KEY-FDBK-AREA.

*    END OF SOURCE CODE
```

Now we can compile and link the program. You'll need to ask your supervisor or teammate for the compile procedure. I am using JCL as follows to execute a COBOL-IMS compile procedure:

```
//USER01D JOB MSGLEVEL=(1,1),NOTIFY=&SYSUID
//*
//* COMPILE A IMS COBOL PROGRAM
//*
//PLIB    JCLLIB ORDER=SYS1.IMS.PROCLIB
//CL      EXEC IMSCOBCL,
//              MBR=COBIMS1,                   <= COBOL PROGRAM NAME
//              SRCLIB=USER01.COBOL.SRCLIB,    <= COBOL SOURCE LIBRARY
//              COPYLIB=USER01.COPYLIB,        <= COPY BOOK LIBRARY
//              LOADLIB=USER01.IMS.LOADLIB     <= LOAD LIBRARY
```

Finally, one time only you must create and use a special PSB to load the database. The PSB can be identical to the one we already created except it must specify a PROCOPT of **LS** which means Load Sequential. Let's clone EMPPSB into member EMPPSBL:

```
PRINT NOGEN
PCB    TYPE=DB,NAME=EMPLOYEE,KEYLEN=20,PROCOPT=LS
SENSEG NAME=EMPLOYEE,PARENT=0
SENSEG NAME=EMPPAY,PARENT=EMPLOYEE
SENSEG NAME=EMPPAYHS,PARENT=EMPPAY
SENSEG NAME=EMPDEP,PARENT=EMPLOYEE
PSBGEN LANG=COBOL,PSBNAME=EMPLOYEE
END
```

Generate this PSB, and then let's execute the program. Execution JCL will look something like this (yours will be whatever you use at your installation). Note that we **MUST** include DD statements for the IMS database and its overflow dataset. Also we include the input file.

```
//USER01D JOB MSGLEVEL=(1,1),NOTIFY=&SYSUID
//*
//* TO RUN A IMS COBOL PROGRAM
//*
//PLIB    JCLLIB ORDER=SYS1.IMS.PROCLIB
//RUN     EXEC IMSCOBGO,
//              MBR=COBIMS1,                   <= COBOL PROGRAM NAME
//              LOADLIB=USER01.IMS.LOADLIB,    <= LOAD LIBRARY
//              PSB=EMPPSBL,          <= PSB NAME
//              PSBLIB=USER01.IMS.PSBLIB,     <= PSB LIBRARY
//              DBDLIB=USER01.IMS.DBDLIB      <= DBD LIBRARY
//*
//** FLAT FILES IF ANY  **********************
//GO.EMPIFILE DD DSN=USER01.EMPIFILE,DISP=SHR
//*
```

```
//** IMS DATABASES (VSAM) ********************
//GO.EMPLOYEE DD DSN=USER01.IMS.EMPLOYEE.CLUSTER,DISP=SHR
//GO.EMPLFLW DD DSN=USER01.IMS.EMPLFLW.CLUSTER,DISP=SHR
//GO.SYSPRINT DD SYSOUT=*
//GO.SYSUDUMP DD SYSOUT=*
//GO.PLIDUMP DD SYSOUT=*
```

And here are the results of the run:

```
** PROGRAM COBIMS1 START **
PROCESSING IN P100-INITIALIZATION
PROCESSING IN P200-MAINLINE
SUCCESSFUL INSERT-REC:1111 VEREEN                CHARLES
SUCCESSFUL INSERT-REC:1122 JENKINS               DEBORAH
SUCCESSFUL INSERT-REC:3217 JOHNSON               EDWARD
SUCCESSFUL INSERT-REC:4175 TURNBULL              FRED
SUCCESSFUL INSERT-REC:4720 SCHULTZ               TIM
SUCCESSFUL INSERT-REC:4836 SMITH                 SANDRA
SUCCESSFUL INSERT-REC:6288 WILLARD               JOE
SUCCESSFUL INSERT-REC:7459 STEWART               BETTY
SUCCESSFUL INSERT-REC:9134 FRANKLIN              BRIANNA
PROCESSING IN P300-TERMINATION
** COBIMS1 - SUCCESSFULLY ENDED **
```

You can browse the IMS data using whatever tool you have such as File Manager IMS.
Or you can simply browse the DATA file of the VSAM data set.

```
Browse            USER01.IMS.EMPLOYEE.DATA                    Top of 9
Command ===>                                                 Scroll PAGE
                      Type DATA       RBA                    Format CHAR
                                         Col 1
----+----10---+----2----+----3----+----4----+----5----+----6----+----7----+----
****  Top of data   ****
......1111 VEREEN                  CHARLES             12 2017-01-01 93
......1122 JENKINS                 DEBORAH             05 2017-01-01 43
......3217 JOHNSON                 EDWARD              04 2017-01-01 39
......4175 TURNBULL                FRED                01 2016-12-01 54
......4720 SCHULTZ                 TIM                 09 2017-01-01 65
......4836 SMITH                   SANDRA              03 2017-01-01 02
......6288 WILLARD                 JOE                 06 2016-01-01 20
......7459 STEWART                 BETTY               07 2016-07-31 01
......9134 FRANKLIN                BRIANNA             00 2016-10-01 93
****  End of data   ****
```

For those of you who code in PLI, we must create a different PSB. All you need to do is clone the EMPPSB we are using for COBOL, and on the last line specify that the language is PL/I instead of COBOL.

```
PRINT NOGEN
PCB    TYPE=DB,NAME=EMPLOYEE,KEYLEN=20,PROCOPT=AP
SENSEG NAME=EMPLOYEE,PARENT=0
SENSEG NAME=EMPPAY,PARENT=EMPLOYEE
SENSEG NAME=EMPPAYHS,PARENT=EMPPAY
SENSEG NAME=EMPDEP,PARENT=EMPLOYEE
PSBGEN LANG=PL/I,PSBNAME=EMPLOYEE
END
```

In my case, I created the PLI PSB with member name EMPPSBP so that it won't conflict with the COBOL PSB named EMPPSB. By the way, if you do try to run a PLI program using a COBOL PSB (or vice-versa) you'll get a U0476 abend. Programs must use a PSB that was generated for that particular programming language.

Now, here is the PLI version of the program to load the IMS database.

```
PLIIMS1: PROCEDURE (DB_PTR_PCB) OPTIONS(MAIN);
/****************************************************************
* PROGRAM NAME :    PLIIMS1 - INSERT RECORDS INTO IMS EMPLOYEE DB    *
****************************************************************/

/****************************************************************
/*                 F I L E S   U S E D                         *
****************************************************************/

   DCL EMPIFILE FILE RECORD SEQL INPUT;

/****************************************************************
/*              W O R K I N G   S T O R A G E                  *
****************************************************************/

   DCL SW_END_OF_FILE         STATIC BIT(01) INIT('0'B);
   DCL ONCODE                 BUILTIN;
   DCL DB_PTR_PCB             POINTER;

   DCL PLITDLI                EXTERNAL ENTRY;

   DCL 01 DLI_FUNCTIONS,
          05 DLI_FUNCISRT     CHAR(04) INIT ('ISRT'),
          05 DLI_FUNCGU       CHAR(04) INIT ('GU  '),
          05 DLI_FUNCGN       CHAR(04) INIT ('GN  '),
```

```
             05  DLI_FUNCGHU          CHAR(04) INIT ('GHU '),
             05  DLI_FUNCGNP          CHAR(04) INIT ('GNP '),
             05  DLI_FUNCREPL         CHAR(04) INIT ('REPL'),
             05  DLI_FUNCDLET         CHAR(04) INIT ('DLET'),
             05  DLI_FUNCXRST         CHAR(04) INIT ('XRST'),
             05  DLI_FUNCCHKP         CHAR(04) INIT ('CHKP'),
             05  DLI_FUNCROLL         CHAR(04) INIT ('ROLL');

DCL 01 IN_EMPLOYEE_RECORD,
             05   EMPL_ID_IN          CHAR(04),
             05   FILLER1             CHAR(01),
             05   EMPL_LNAME          CHAR(30),
             05   FILLER2             CHAR(01),
             05   EMPL_FNAME          CHAR(20),
             05   FILLER3             CHAR(01),
             05   EMPL_YRS_SRV        CHAR(02),
             05   FILLER4             CHAR(01),
             05   EMPL_PRM_DTE        CHAR(10),
             05   FILLER5             CHAR(10);

DCL 01 PCB_MASK              BASED(DB_PTR_PCB),
             05 DBD_NAME              CHAR(08),
             05 SEG_LEVEL             CHAR(02),
             05 STATUS_CODE           CHAR(02),
             05 PROC_OPT              CHAR(04),
             05 FILLER6               FIXED BIN (31),
             05 SEG_NAME              CHAR(08),
             05 KEY_FDBK              FIXED BIN (31),
             05 NUM_SENSEG            FIXED BIN (31),
             05 KEY_FDBK_AREA,
                10 EMPLOYEE_ID        CHAR(04);

DCL 01 EMP_UNQUALIFIED_SSA,
             05   SEGNAME             CHAR(08) INIT ('EMPLOYEE'),
             05   FILLER7             CHAR(01) INIT (' ');

DCL 01 EMP_QUALIFIED_SSA,
             05   SEGNAME             CHAR(08) INIT('EMPLOYEE'),
             05   FILLER8             CHAR(01) INIT('('),
             05   FIELD               CHAR(08) INIT('EMPID'),
             05   OPER                CHAR(02) INIT(' ='),
             05   EMP_ID_VAL          CHAR(04) INIT('    '),
             05   FILLER9             CHAR(01) INIT(')');

DCL SEG_IO_AREA              CHAR(80) INIT (' ');

DCL THREE                    FIXED BIN (31) INIT(3);
DCL FOUR                     FIXED BIN (31) INIT(4);
```

```
   DCL FIVE                       FIXED BIN (31) INIT(5);
   DCL SIX                        FIXED BIN (31) INIT(6);

/*******************************************************************
/*              O N   C O N D I T I O N S                        *
*******************************************************************/

   ON ENDFILE (EMPIFILE) SW_END_OF_FILE =  '1'B;

/*******************************************************************
/*              P R O G R A M   M A I N L I N E                  *
*******************************************************************/

CALL P100_INITIALIZATION;
CALL P200_MAINLINE;
CALL P300_TERMINATION;

P100_INITIALIZATION: PROC;

   PUT SKIP LIST ('PLIIMS1: INSERT RECORDS');
   OPEN FILE (EMPIFILE);

   IN_EMPLOYEE_RECORD  = '';
   PCB_MASK = '';

END P100_INITIALIZATION;

P200_MAINLINE: PROC;

   /*  MAIN LOOP _ READ THE INPUT FILE, LOAD THE OUTPUT
                   STRUCTURE AND WRITE THE RECORD TO OUTPUT */

   READ FILE (EMPIFILE) INTO (IN_EMPLOYEE_RECORD);

   DO WHILE (¬SW_END_OF_FILE);

      SEG_IO_AREA = STRING(IN_EMPLOYEE_RECORD);

      CALL PLITDLI (FOUR,
                    DLI_FUNCISRT,
                    PCB_MASK,
                    SEG_IO_AREA,
                    EMP_UNQUALIFIED_SSA);

      IF STATUS_CODE = '  ' THEN
         PUT SKIP LIST ('SUCCESSFUL INSERT-REC:' || SEG_IO_AREA);
      ELSE
         DO;
```

41

```
                    CALL P400_DISPLAY_ERROR;
                    RETURN;
                END;

            READ FILE (EMPIFILE) INTO (IN_EMPLOYEE_RECORD);

        END; /* DO WHILE */

END P200_MAINLINE;

P300_TERMINATION: PROC;

    CLOSE FILE(EMPIFILE);

    PUT SKIP LIST ('PLIIMS1 - SUCCESSFULLY ENDED');

END P300_TERMINATION;

P400_DISPLAY_ERROR: PROC;

    PUT SKIP LIST ('ERROR ENCOUNTERED - DETAIL FOLLOWS');
    PUT SKIP LIST ('SEG_IO_AREA    :' || SEG_IO_AREA);
    PUT SKIP LIST ('DBD_NAME1:' || DBD_NAME);
    PUT SKIP LIST ('SEG_LEVEL1:' || SEG_LEVEL);
    PUT SKIP LIST ('STATUS_CODE:' || STATUS_CODE);
    PUT SKIP LIST ('PROC_OPT1 :' || PROC_OPT);
    PUT SKIP LIST ('SEG_NAME1 :' || SEG_NAME);
    PUT SKIP LIST ('KEY_FDBK1 :' || KEY_FDBK);
    PUT SKIP LIST ('NUM_SENSEG1:' || NUM_SENSEG);
    PUT SKIP LIST ('KEY_FDBK_AREA1:' || KEY_FDBK_AREA);

END P400_DISPLAY_ERROR;

END PLIIMS1;
```

Reading an IMS Segment

Our next program will be named COBIMS2, and the purpose is simply to retrieve a record from the EMPLOYEE database. In this case, we want the record for employee 3217.

Our basic program structure will be similar to the load program except we will need to perform a Get Unique (GU) call, and we'll use a qualified SSA. Remember our qualified SSA structure looks like this:

```
01 EMP-QUALIFIED-SSA.
   05  SEGNAME     PIC X(08) VALUE 'EMPLOYEE'.
   05  FILLER      PIC X(01) VALUE '('.
   05  FIELD       PIC X(08) VALUE 'EMPID'.
   05  OPER        PIC X(02) VALUE ' ='.
   05  EMP-ID-VAL  PIC X(04) VALUE '    '.
   05  FILLER      PIC X(01) VALUE ')'.
```

So we must load the EMP-ID-VAL variable with character value '3217'. Our call will look like this.

```
CALL 'CBLTDLI' USING FOUR,
               DLI-FUNCGU,
               PCB-MASK,
               SEG-IO-AREA,
               EMP-QUALIFIED-SSA
```

Now we can code the entire program. We don't need a loop because we are retrieving a single record. So the program is quite simple. Note that we check for a blank status code after the IMS call, and we report an error if it is not blank.

```
ID DIVISION.
PROGRAM-ID. COBIMS2.
*****************************************************
*    RETRIEVE A RECORD FROM IMS EMPLOYEE DATABASE    *
*****************************************************

ENVIRONMENT DIVISION.
DATA DIVISION.

*****************************************************
*  W O R K I N G   S T O R A G E   S E C T I O N    *
*****************************************************

WORKING-STORAGE SECTION.

01 SEG-IO-AREA     PIC X(80).

01 DLI-FUNCTIONS.
   05 DLI-FUNCISRT PIC X(4) VALUE 'ISRT'.
   05 DLI-FUNCGU   PIC X(4) VALUE 'GU  '.
   05 DLI-FUNCGN   PIC X(4) VALUE 'GN  '.
   05 DLI-FUNCGHU  PIC X(4) VALUE 'GHU '.
   05 DLI-FUNCGNP  PIC X(4) VALUE 'GNP '.
   05 DLI-FUNCREPL PIC X(4) VALUE 'REPL'.
   05 DLI-FUNCDLET PIC X(4) VALUE 'DLET'.
```

43

```
       05 DLI-FUNCXRST  PIC X(4) VALUE 'XRST'.
       05 DLI-FUNCCKPT  PIC X(4) VALUE 'CKPT'.


   01 EMP-UNQUALIFIED-SSA.
       05  SEGNAME     PIC X(08) VALUE 'EMPLOYEE'.
       05  FILLER      PIC X(01) VALUE ' '.


   01 EMP-QUALIFIED-SSA.
       05  SEGNAME     PIC X(08) VALUE 'EMPLOYEE'.
       05  FILLER      PIC X(01) VALUE '('.
       05  FIELD       PIC X(08) VALUE 'EMPID'.
       05  OPER        PIC X(02) VALUE ' ='.
       05  EMP-ID-VAL  PIC X(04) VALUE '    '.
       05  FILLER      PIC X(01) VALUE ')'.


   01 IMS-RET-CODES.
       05 THREE        PIC S9(9) COMP VALUE +3.
       05 FOUR         PIC S9(9) COMP VALUE +4.
       05 FIVE         PIC S9(9) COMP VALUE +5.
       05 SIX          PIC S9(9) COMP VALUE +6.


   LINKAGE SECTION.
   01 PCB-MASK.
       03 DBD-NAME     PIC X(8).
       03 SEG-LEVEL    PIC XX.
       03 STATUS-CODE  PIC XX.
       03 PROC-OPT     PIC X(4).
       03 FILLER       PIC X(4).
       03 SEG-NAME     PIC X(8).
       03 KEY-FDBK     PIC S9(5) COMP.
       03 NUM-SENSEG   PIC S9(5) COMP.
       03 KEY-FDBK-AREA.
          05 EMPLOYEE-ID  PIC X(04).
          05 EMPPAYHS     PIC X(08).


   PROCEDURE DIVISION.

       INITIALIZE PCB-MASK
       ENTRY 'DLITCBL' USING PCB-MASK

       PERFORM P100-INITIALIZATION.
       PERFORM P200-MAINLINE.
       PERFORM P300-TERMINATION.
       GOBACK.
```

```
P100-INITIALIZATION.

    DISPLAY '** PROGRAM COBIMS2 START **'
    DISPLAY 'PROCESSING IN P100-INITIALIZATION'.


P200-MAINLINE.

    DISPLAY 'PROCESSING IN P200-MAINLINE'

    MOVE '3217' TO EMP-ID-VAL

    CALL 'CBLTDLI' USING FOUR,
                  DLI-FUNCGU,
                  PCB-MASK,
                  SEG-IO-AREA,
                  EMP-QUALIFIED-SSA

    IF STATUS-CODE = '  '
       DISPLAY 'SUCCESSFUL GET CALL  '
       DISPLAY 'SEG-IO-ARE : ' SEG-IO-AREA
    ELSE
       DISPLAY 'ERROR IN FETCH :' STATUS-CODE
       PERFORM P400-DISPLAY-ERROR
    END-IF.


P300-TERMINATION.

    DISPLAY 'PROCESSING IN P300-TERMINATION'
    DISPLAY '** COBIMS2 - SUCCESSFULLY ENDED **'.


P400-DISPLAY-ERROR.

    DISPLAY 'ERROR ENCOUNTERED - DETAIL FOLLOWS'
    DISPLAY 'SEG-IO-AREA      :' SEG-IO-AREA
    DISPLAY 'DBD-NAME1:'      DBD-NAME
    DISPLAY 'SEG-LEVEL1:'     SEG-LEVEL
    DISPLAY 'STATUS-CODE:'    STATUS-CODE
    DISPLAY 'PROC-OPT1 :'     PROC-OPT
    DISPLAY 'SEG-NAME1 :'     SEG-NAME
    DISPLAY 'KEY-FDBK1 :'     KEY-FDBK
    DISPLAY 'NUM-SENSEG1:'    NUM-SENSEG
    DISPLAY 'KEY-FDBK-AREA1:' KEY-FDBK-AREA.

*    END OF SOURCE CODE
```

Now compile, link and run the program. Here is the output showing that the data was successfully retrieved.

```
** PROGRAM COBIMS2 START **
PROCESSING IN P100-INITIALIZATION
PROCESSING IN P200-MAINLINE
SUCCESSFUL GET CALL
SEG-IO-ARE : 3217 JOHNSON             EDWARD              04 2017-01-01
397342007
PROCESSING IN P300-TERMINATION

** COBIMS2 - SUCCESSFULLY ENDED **
```

Also, we need to test a case where we try to retrieve an employee number which doesn't exist. Let's modify the program to look for EMP-ID 3218 which doesn't exist. Now recompile and re-execute the program. Here's the result:

```
** PROGRAM COBIMS2 START **
PROCESSING IN P100-INITIALIZATION
PROCESSING IN P200-MAINLINE
ERROR IN FETCH :GE
ERROR ENCOUNTERED - DETAIL FOLLOWS
SEG-IO-AREA       :
DBD-NAME1:EMPLOYEE
SEG-LEVEL1:00
STATUS-CODE:GE
PROC-OPT1 :AP
SEG-NAME1 :
KEY-FDBK1 :00000
NUM-SENSEG1:00004
KEY-FDBK-AREA1:
PROCESSING IN P300-TERMINATION
** COBIMS2 - SUCCESSFULLY ENDED **
```

Excellent, we captured and reported the error. IMS returned a GE return code which means the record was not found.

You'll use GU processing anytime you have a need to access data for a particular record in the database for read-only. Here we read all the root segments. Later we will read segments lower in the database hierarchy. Here is the PLI source code for the same program.

```
PLIIMS2: PROCEDURE (DB_PTR_PCB) OPTIONS(MAIN);
/*************************************************************
 * PROGRAM NAME :   PLIIMS2 - RETRIEVE A RECORD FROM EMPLOYEE DB    *
 *************************************************************/
```

```
/*******************************************************************
/*                W O R K I N G   S T O R A G E                    *
********************************************************************/

    DCL ONCODE                  BUILTIN;
    DCL DB_PTR_PCB              POINTER;
    DCL PLITDLI                 EXTERNAL ENTRY;

    DCL 01 DLI_FUNCTIONS,
         05 DLI_FUNCISRT        CHAR(04) INIT ('ISRT'),
         05 DLI_FUNCGU          CHAR(04) INIT ('GU  '),
         05 DLI_FUNCGN          CHAR(04) INIT ('GN  '),
         05 DLI_FUNCGHU         CHAR(04) INIT ('GHU '),
         05 DLI_FUNCGNP         CHAR(04) INIT ('GNP '),
         05 DLI_FUNCREPL        CHAR(04) INIT ('REPL'),
         05 DLI_FUNCDLET        CHAR(04) INIT ('DLET'),
         05 DLI_FUNCXRST        CHAR(04) INIT ('XRST'),
         05 DLI_FUNCCHKP        CHAR(04) INIT ('CHKP'),
         05 DLI_FUNCROLL        CHAR(04) INIT ('ROLL');

    DCL 01 IO_EMPLOYEE_RECORD,
         05   EMPL_ID_IN        CHAR(04),
         05   FILLER1           CHAR(01),
         05   EMPL_LNAME        CHAR(30),
         05   FILLER2           CHAR(01),
         05   EMPL_FNAME        CHAR(20),
         05   FILLER3           CHAR(01),
         05   EMPL_YRS_SRV      CHAR(02),
         05   FILLER4           CHAR(01),
         05   EMPL_PRM_DTE      CHAR(10),
         05   FILLER5           CHAR(10);

    DCL 01 PCB_MASK             BASED(DB_PTR_PCB),
         05 DBD_NAME            CHAR(08),
         05 SEG_LEVEL           CHAR(02),
         05 STATUS_CODE         CHAR(02),
         05 PROC_OPT            CHAR(04),
         05 FILLER6             FIXED BIN (31),
         05 SEG_NAME            CHAR(08),
         05 KEY_FDBK            FIXED BIN (31),
         05 NUM_SENSEG          FIXED BIN (31),
         05 KEY_FDBK_AREA,
            10 EMPLOYEE_ID      CHAR(04);

    DCL 01 EMP_UNQUALIFIED_SSA,
         05   SEGNAME           CHAR(08) INIT ('EMPLOYEE'),
         05   FILLER7           CHAR(01) INIT (' ');
```

```
      DCL 01 EMP_QUALIFIED_SSA,
              05   SEGNAME          CHAR(08)  INIT('EMPLOYEE'),
              05   FILLER8          CHAR(01)  INIT('('),
              05   FIELD            CHAR(08)  INIT('EMPID'),
              05   OPER             CHAR(02)  INIT(' ='),
              05   EMP_ID_VAL       CHAR(04)  INIT('    '),
              05   FILLER9          CHAR(01)  INIT(')');

      DCL THREE                     FIXED BIN (31) INIT(3);
      DCL FOUR                      FIXED BIN (31) INIT(4);
      DCL FIVE                      FIXED BIN (31) INIT(5);
      DCL SIX                       FIXED BIN (31) INIT(6);

/*****************************************************************
/*                  P R O G R A M   M A I N L I N E            *
*****************************************************************/
CALL P100_INITIALIZATION;
CALL P200_MAINLINE;
CALL P300_TERMINATION;

P100_INITIALIZATION: PROC;

    PUT SKIP LIST ('PLIIMS2: RETRIEVE RECORD FROM EMPLOYEE DB');
    IO_EMPLOYEE_RECORD  = '';

END P100_INITIALIZATION;

P200_MAINLINE: PROC;

   /*  SET THE EMPLOYEE SEGMENT SEARCH ARGUMENT AND CALL PLITDLI */

   EMP_ID_VAL = '3217';

   CALL PLITDLI (FOUR,
                 DLI_FUNCGU,
                 PCB_MASK,
                 IO_EMPLOYEE_RECORD,
                 EMP_QUALIFIED_SSA);

   IF STATUS_CODE = '  ' THEN
      DO;
         PUT SKIP LIST ('SUCCESSFUL RETRIEVAL :');
         PUT SKIP DATA(IO_EMPLOYEE_RECORD);
      END;
   ELSE
      CALL P400_DISPLAY_ERROR;

END P200_MAINLINE;
```

```
P300_TERMINATION: PROC;

    PUT SKIP LIST ('PLIIMS2 - ENDED SUCCESSFULLY');

END P300_TERMINATION;

P400_DISPLAY_ERROR: PROC;

    PUT SKIP LIST ('ERROR ENCOUNTERED - DETAIL FOLLOWS');
    PUT SKIP LIST ('SEG_IO_AREA    :' || SEG_IO_AREA);
    PUT SKIP LIST ('DBD_NAME1:' || DBD_NAME);
    PUT SKIP LIST ('SEG_LEVEL1:' || SEG_LEVEL);
    PUT SKIP LIST ('STATUS_CODE:' || STATUS_CODE);
    PUT SKIP LIST ('PROC_OPT1 :' || PROC_OPT);
    PUT SKIP LIST ('SEG_NAME1 :' || SEG_NAME);
    PUT SKIP LIST ('KEY_FDBK1 :' || KEY_FDBK);
    PUT SKIP LIST ('NUM_SENSEG1:' || NUM_SENSEG);
    PUT SKIP LIST ('KEY_FDBK_AREA1:' || KEY_FDBK_AREA);

END P400_DISPLAY_ERROR;

END PLIIMS2;
```

And here is the output for retrieving employee 3217.

```
PLIIMS2: RETRIEVE RECORD FROM EMPLOYEE DB
SUCCESSFUL RETRIEVAL :
IO_EMPLOYEE_RECORD.EMPL_ID_IN='3217'          IO_EMPLOYEE_RECORD.FILLER1=' '
IO_EMPLOYEE_RECORD.EMPL_LNAME='JOHNSON              '          IO_EMPLO
IO_EMPLOYEE_RECORD.EMPL_FNAME='EDWARD            '    '          IO_EMPLO
IO_EMPLOYEE_RECORD.EMPL_YRS_SRV='04'          IO_EMPLOYEE_RECORD.FILLER4=' '
IO_EMPLOYEE_RECORD.EMPL_PRM_DTE='2017-01-01'  IO_EMPLOYEE_RECORD.FILLER5=' 397
PLIIMS2 - ENDED SUCCESSFULLY
```

If we change the record number to employee 3218, we get this output:

```
PLIIMS2: RETRIEVE RECORD FROM EMPLOYEE DB
ERROR ENCOUNTERED - DETAIL FOLLOWS
SEG_IO_AREA    :
DBD_NAME1:EMPLOYEE
SEG_LEVEL1:00
STATUS_CODE:GE
PROC_OPT1 :AP
SEG_NAME1 :
KEY_FDBK1 :              0
NUM_SENSEG1:             4
KEY_FDBK_AREA1:
PLIIMS2 - ENDED SUCCESSFULLY
```

Reading a Database Sequentially (GN)

Our next program COBIMS3 will read the entire database sequentially. This scenario isn't unusual (a payroll program might process the database sequentially to generate pay checks) so you'll want to have a model of how to carry it out.

Basically we are going to create a loop that will walk through the database sequentially getting each EMPLOYEE segment using Get Next (GN) calls. We'll need a switch to indicate a stopping point which will be the end of the database (IMS status code GB). We'll also use an unqualified SSA since we don't need to know the key of each record to traverse the database. Here is the code.

```
        IDENTIFICATION DIVISION.
        PROGRAM-ID. COBIMS3.
        ****************************************************
        *   WALK THROUGH THE EMPLOYEE (ROOT) SEGMENTS OF    *
        *   THE ENTIRE EMPLOYEE IMS DATABASE.               *
        ****************************************************

        ENVIRONMENT DIVISION.
        INPUT-OUTPUT SECTION.
        DATA DIVISION.


        ****************************************************
        *   W O R K I N G   S T O R A G E   S E C T I O N   *
        ****************************************************
        WORKING-STORAGE SECTION.

        01 WS-FLAGS.
            05  SW-END-OF-DB-SWITCH    PIC X(1) VALUE 'N'.
                88  SW-END-OF-DB                VALUE 'Y'.
                88  SW-NOT-END-OF-DB            VALUE 'N'.

        01 DLI-FUNCTIONS.
            05 DLI-FUNCISRT  PIC X(4) VALUE 'ISRT'.
            05 DLI-FUNCGU    PIC X(4) VALUE 'GU  '.
            05 DLI-FUNCGN    PIC X(4) VALUE 'GN  '.
            05 DLI-FUNCGHU   PIC X(4) VALUE 'GHU '.
            05 DLI-FUNCGNP   PIC X(4) VALUE 'GNP '.
            05 DLI-FUNCREPL  PIC X(4) VALUE 'REPL'.
            05 DLI-FUNCDLET  PIC X(4) VALUE 'DLET'.
            05 DLI-FUNCXRST  PIC X(4) VALUE 'XRST'.
            05 DLI-FUNCCKPT  PIC X(4) VALUE 'CKPT'.

        01 IN-EMPLOYEE-RECORD.
```

```
       05  EMPL-ID-IN    PIC X(04).
       05  FILLER        PIC X(01).
       05  EMPL-LNAME    PIC X(30).
       05  FILLER        PIC X(01).
       05  EMPL-FNAME    PIC X(20).
       05  FILLER        PIC X(01).
       05  EMPL-YRS-SRV  PIC X(02).
       05  FILLER        PIC X(01).
       05  EMPL-PRM-DTE  PIC X(10).
       05  FILLER        PIC X(10).


   01 EMP-UNQUALIFIED-SSA.
       05  SEGNAME       PIC X(08) VALUE 'EMPLOYEE'.
       05  FILLER        PIC X(01) VALUE ' '.


   01 EMP-QUALIFIED-SSA.
       05  SEGNAME       PIC X(08) VALUE 'EMPLOYEE'.
       05  FILLER        PIC X(01) VALUE '('.
       05  FIELD         PIC X(08) VALUE 'EMPID'.
       05  OPER          PIC X(02) VALUE ' ='.
       05  EMP-ID-VAL    PIC X(04) VALUE '    '.
       05  FILLER        PIC X(01) VALUE ')'.


 01 SEG-IO-AREA     PIC X(80).

 01 IMS-RET-CODES.
     05 THREE           PIC S9(9) COMP VALUE +3.
     05 FOUR            PIC S9(9) COMP VALUE +4.
     05 FIVE            PIC S9(9) COMP VALUE +5.
     05 SIX             PIC S9(9) COMP VALUE +6.

LINKAGE SECTION.
 01 PCB-MASK.
     03 DBD-NAME        PIC X(8).
     03 SEG-LEVEL       PIC XX.
     03 STATUS-CODE     PIC XX.
     03 PROC-OPT        PIC X(4).
     03 FILLER          PIC X(4).
     03 SEG-NAME        PIC X(8).
     03 KEY-FDBK        PIC S9(5) COMP.
     03 NUM-SENSEG      PIC S9(5) COMP.
     03 KEY-FDBK-AREA.
        05 EMPLOYEE-KEY  PIC X(04).
        05 EMPPAYHS-KEY  PIC X(08).

PROCEDURE DIVISION.

    INITIALIZE PCB-MASK
```

```
            ENTRY 'DLITCBL' USING PCB-MASK

            PERFORM P100-INITIALIZATION.
            PERFORM P200-MAINLINE.
            PERFORM P300-TERMINATION.
            GOBACK.

        P100-INITIALIZATION.

            DISPLAY '** PROGRAM COBIMS3 START **'
            DISPLAY 'PROCESSING IN P100-INITIALIZATION'.

*       DO INITIAL DB READ FOR FIRST EMPLOYEE RECORD

            CALL 'CBLTDLI' USING FOUR,
                 DLI-FUNCGN,
                 PCB-MASK,
                 SEG-IO-AREA,
                 EMP-UNQUALIFIED-SSA

            IF STATUS-CODE = '  ' THEN
               NEXT SENTENCE
            ELSE
               IF STATUS-CODE = 'GB' THEN
                  SET SW-END-OF-DB TO TRUE
                  DISPLAY 'END OF DATABASE :'
               ELSE
                  PERFORM P400-DISPLAY-ERROR
                  GOBACK
               END-IF

            END-IF.

        P200-MAINLINE.

            DISPLAY 'PROCESSING IN P200-MAINLINE'

*       CHECK STATUS CODE AND FIRST RECORD

            IF SW-END-OF-DB THEN
               DISPLAY 'NO RECORDS TO PROCESS!!'
            ELSE
               PERFORM UNTIL SW-END-OF-DB
                  DISPLAY 'SUCCESSFUL READ :' SEG-IO-AREA

                  CALL 'CBLTDLI' USING FOUR,
                       DLI-FUNCGN,
                       PCB-MASK,
```

```
                SEG-IO-AREA,
                EMP-UNQUALIFIED-SSA

        IF STATUS-CODE = 'GB' THEN
            SET SW-END-OF-DB TO TRUE
            DISPLAY 'END OF DATABASE'
        ELSE
            IF STATUS-CODE NOT EQUAL SPACES THEN
                PERFORM P400-DISPLAY-ERROR
                GOBACK
            END-IF
        END-IF

    END-PERFORM.

        DISPLAY 'FINISHED PROCESSING IN P200-MAINLINE'.

    P300-TERMINATION.

        DISPLAY 'PROCESSING IN P300-TERMINATION'
        DISPLAY '** COBIMS3 - SUCCESSFULLY ENDED **'.

    P400-DISPLAY-ERROR.

        DISPLAY 'ERROR ENCOUNTERED - DETAIL FOLLOWS'
        DISPLAY 'SEG-IO-AREA      :' SEG-IO-AREA
        DISPLAY 'DBD-NAME1:'       DBD-NAME
        DISPLAY 'SEG-LEVEL1:'      SEG-LEVEL
        DISPLAY 'STATUS-CODE:'     STATUS-CODE
        DISPLAY 'PROC-OPT1 :'      PROC-OPT
        DISPLAY 'SEG-NAME1 :'      SEG-NAME
        DISPLAY 'KEY-FDBK1 :'      KEY-FDBK
        DISPLAY 'NUM-SENSEG1:'     NUM-SENSEG
        DISPLAY 'KEY-FDBK-AREA1:' KEY-FDBK-AREA.

    *    END OF SOURCE CODE
```

Now let's compile and link, and then execute COBIMS3. Here's the run output.

```
** PROGRAM COBIMS3 START **
PROCESSING IN P100-INITIALIZATION
PROCESSING IN P200-MAINLINE
SUCCESSFUL READ :1111 VEREEN          CHARLES        12 201
SUCCESSFUL READ :1122 JENKINS         DEBORAH        05 201
SUCCESSFUL READ :3217 JOHNSON         EDWARD         04 201
SUCCESSFUL READ :4175 TURNBULL        FRED           01 201
SUCCESSFUL READ :4720 SCHULTZ         TIM            09 201
SUCCESSFUL READ :4836 SMITH           SANDRA         03 201
```

```
SUCCESSFUL READ :6288 WILLARD              JOE              06 201
SUCCESSFUL READ :7459 STEWART              BETTY            07 201
SUCCESSFUL READ :9134 FRANKLIN             BRIANNA          00 201
END OF DATABASE
FINISHED PROCESSING IN P200-MAINLINE
PROCESSING IN P300-TERMINATION
** COBIMS3 - SUCCESSFULLY ENDED **
```

You now have a model for any kind of sequential processing you want to do on root segments. Processing child segments is a bit more involved, but not much. We'll show an example of that later.

Here's the PLI source for the program:

```
PLIIMS3: PROCEDURE (DB_PTR_PCB) OPTIONS(MAIN);
 /******************************************************************
 * PROGRAM NAME: PLIIMS3 - WALK THROUGH THE EMPLOYEE (ROOT)       *
 *                         SEGMENTS OF THE EMPLOYEE IMS DATABASE.  *
 ******************************************************************/
 /******************************************************************
 /*                 W O R K I N G   S T O R A G E                 *
 ******************************************************************/

    DCL SW_END_OF_DB            STATIC BIT(01) INIT('0'B);
    DCL ONCODE                  BUILTIN;
    DCL DB_PTR_PCB              POINTER;

    DCL PLITDLI                 EXTERNAL ENTRY;
    DCL 01 DLI_FUNCTIONS,
          05 DLI_FUNCISRT       CHAR(04) INIT ('ISRT'),
          05 DLI_FUNCGU         CHAR(04) INIT ('GU  '),
          05 DLI_FUNCGN         CHAR(04) INIT ('GN  '),
          05 DLI_FUNCGHU        CHAR(04) INIT ('GHU '),
          05 DLI_FUNCGNP        CHAR(04) INIT ('GNP '),
          05 DLI_FUNCREPL       CHAR(04) INIT ('REPL'),
          05 DLI_FUNCDLET       CHAR(04) INIT ('DLET'),
          05 DLI_FUNCXRST       CHAR(04) INIT ('XRST'),
          05 DLI_FUNCCHKP       CHAR(04) INIT ('CHKP'),
          05 DLI_FUNCROLL       CHAR(04) INIT ('ROLL');

    DCL 01 IO_EMPLOYEE_RECORD,
          05  EMPL_ID_IN        CHAR(04),
          05  FILLER1           CHAR(01),
          05  EMPL_LNAME        CHAR(30),
          05  FILLER2           CHAR(01),
          05  EMPL_FNAME        CHAR(20),
          05  FILLER3           CHAR(01),
          05  EMPL_YRS_SRV      CHAR(02),
          05  FILLER4           CHAR(01),
          05  EMPL_PRM_DTE      CHAR(10),
          05  FILLER5           CHAR(10);
```

```
DCL 01 PCB_MASK               BASED(DB_PTR_PCB),
        05 DBD_NAME           CHAR(08),
        05 SEG_LEVEL          CHAR(02),
        05 STATUS_CODE        CHAR(02),
        05 PROC_OPT           CHAR(04),
        05 FILLER6            FIXED BIN (31),
        05 SEG_NAME           CHAR(08),
        05 KEY_FDBK           FIXED BIN (31),
        05 NUM_SENSEG         FIXED BIN (31),
        05 KEY_FDBK_AREA,
           10 EMPLOYEE_ID     CHAR(04);


DCL 01 EMP_UNQUALIFIED_SSA,
        05  SEGNAME           CHAR(08) INIT ('EMPLOYEE'),
        05  FILLER7           CHAR(01) INIT (' ');

DCL 01 EMP_QUALIFIED_SSA,
        05  SEGNAME           CHAR(08) INIT('EMPLOYEE'),
        05  FILLER8           CHAR(01) INIT('('),
        05  FIELD             CHAR(08) INIT('EMPID'),
        05  OPER              CHAR(02) INIT(' ='),
        05  EMP_ID_VAL        CHAR(04) INIT('    '),
        05  FILLER9           CHAR(01) INIT(')');

DCL THREE                     FIXED BIN (31) INIT(3);
DCL FOUR                      FIXED BIN (31) INIT(4);
DCL FIVE                      FIXED BIN (31) INIT(5);
DCL SIX                       FIXED BIN (31) INIT(6);

/********************************************************************
/*              P R O G R A M   M A I N L I N E              *
********************************************************************/
CALL P100_INITIALIZATION;
CALL P200_MAINLINE;
CALL P300_TERMINATION;

P100_INITIALIZATION: PROC;

    PUT SKIP LIST ('PLIIMS3: TRAVERSE EMPLOYEE DATABASE ROOT SEGS');
    PCB_MASK = '';
    IO_EMPLOYEE_RECORD  = '';

 /* DO INITIAL DB READ FOR FIRST EMPLOYEE RECORD */

    CALL PLITDLI (FOUR,
                  DLI_FUNCGN,
                  PCB_MASK,
                  IO_EMPLOYEE_RECORD,
                  EMP_UNQUALIFIED_SSA);

    IF STATUS_CODE = '  ' THEN;
    ELSE
```

55

```
               IF STATUS_CODE = 'GB' THEN
                   DO;
                        SW_END_OF_DB = '1'B;
                        PUT SKIP LIST ('** END OF DATABASE');
                   END;
               ELSE
                   DO;
                        CALL P400_DISPLAY_ERROR;
                        RETURN;
                   END;

      END P100_INITIALIZATION;

      P200_MAINLINE: PROC;

          /*  MAIN LOOP - CYCLE THROUGH ALL ROOT SEGMENTS IN THE DB,
                      DISPLAYING THE DATA RETRIEVED                 */

              IF SW_END_OF_DB THEN
                  PUT SKIP LIST ('NO RECORDS TO PROCESS!!');
              ELSE
                  DO WHILE (¬SW_END_OF_DB);
                      PUT SKIP LIST ('SUCCESSFUL READ :'
                          || IO_EMPLOYEE_RECORD);

                          CALL PLITDLI (FOUR,
                                        DLI_FUNCGN,
                                        PCB_MASK,
                                        IO_EMPLOYEE_RECORD,
                                        EMP_UNQUALIFIED_SSA);

                      IF STATUS_CODE = '  ' THEN;
                      ELSE
                          IF STATUS_CODE = 'GB' THEN
                              DO;
                                  SW_END_OF_DB = '1'B;
                                  PUT SKIP LIST ('** END OF DATABASE');
                              END;
                          ELSE
                              DO;
                                  CALL P400_DISPLAY_ERROR;
                                  RETURN;
                              END;

                  END; /* DO WHILE */

              PUT SKIP LIST ('FINISHED PROCESSING IN P200_MAINLINE');

      END P200_MAINLINE;

      P300_TERMINATION: PROC;
          CLOSE FILE(EMPIFILE);

          PUT SKIP LIST ('PLIIMS3 - SUCCESSFULLY ENDED');
```

56

```
      END P300_TERMINATION;

   P400_DISPLAY_ERROR: PROC;

       PUT SKIP LIST ('ERROR ENCOUNTERED - DETAIL FOLLOWS');
       PUT SKIP LIST ('SEG_IO_AREA     :' || SEG_IO_AREA);
       PUT SKIP LIST ('DBD_NAME1:' ||  DBD_NAME);
       PUT SKIP LIST ('SEG_LEVEL1:' || SEG_LEVEL);
       PUT SKIP LIST ('STATUS_CODE:' || STATUS_CODE);
       PUT SKIP LIST ('PROC_OPT1 :' || PROC_OPT);
       PUT SKIP LIST ('SEG_NAME1 :' || SEG_NAME);
       PUT SKIP LIST ('KEY_FDBK1 :' || KEY_FDBK);
       PUT SKIP LIST ('NUM_SENSEG1:' || NUM_SENSEG);
       PUT SKIP LIST ('KEY_FDBK_AREA1:' || KEY_FDBK_AREA);

   END P400_DISPLAY_ERROR;

   END PLIIMS3;
```

The output from PLIIMS3 is here:

```
PLIIMS3: TRAVERSE EMPLOYEE DATABASE ROOT SEGS
SUCCESSFUL READ :1111   SUCCESSFUL READ :      SUCCESSFUL READ :VEREEN
SUCCESSFUL READ :CHARLES                       SUCCESSFUL READ :       SUCCESSF
SUCCESSFUL READ :2017-01-01                     SUCCESSFUL READ : 937253058
SUCCESSFUL READ :1122   SUCCESSFUL READ :      SUCCESSFUL READ :JENKINS
SUCCESSFUL READ :DEBORAH                        SUCCESSFUL READ :       SUCCESSF
SUCCESSFUL READ :2017-01-01                     SUCCESSFUL READ : 435092366
SUCCESSFUL READ :3217   SUCCESSFUL READ :      SUCCESSFUL READ :JOHNSON
SUCCESSFUL READ :EDWARD                         SUCCESSFUL READ :       SUCCESSF
SUCCESSFUL READ :2017-01-01                     SUCCESSFUL READ : 397342007
SUCCESSFUL READ :4175   SUCCESSFUL READ :      SUCCESSFUL READ :TURNBULL
SUCCESSFUL READ :FRED                           SUCCESSFUL READ :       SUCCESSF
SUCCESSFUL READ :2016-12-01                     SUCCESSFUL READ : 542083017
SUCCESSFUL READ :4720   SUCCESSFUL READ :      SUCCESSFUL READ :SCHULTZ
SUCCESSFUL READ :TIM                            SUCCESSFUL READ :       SUCCESSF
SUCCESSFUL READ :2017-01-01                     SUCCESSFUL READ : 650450254
SUCCESSFUL READ :4836   SUCCESSFUL READ :      SUCCESSFUL READ :SMITH
SUCCESSFUL READ :SANDRA                         SUCCESSFUL READ :       SUCCESSF
SUCCESSFUL READ :2017-01-01                     SUCCESSFUL READ : 028374669
SUCCESSFUL READ :6288   SUCCESSFUL READ :      SUCCESSFUL READ :WILLARD
SUCCESSFUL READ :JOE                            SUCCESSFUL READ :       SUCCESSF
SUCCESSFUL READ :2016-01-01                     SUCCESSFUL READ : 209883920
SUCCESSFUL READ :7459   SUCCESSFUL READ :      SUCCESSFUL READ :STEWART
SUCCESSFUL READ :BETTY                          SUCCESSFUL READ :       SUCCESSF
SUCCESSFUL READ :2016-07-31                     SUCCESSFUL READ : 019572830
SUCCESSFUL READ :9134   SUCCESSFUL READ :      SUCCESSFUL READ :FRANKLIN
SUCCESSFUL READ :BRIANNA                        SUCCESSFUL READ :       SUCCESSF
SUCCESSFUL READ :2016-10-01                     SUCCESSFUL READ : 937293598
** END OF DATABASE
FINISHED PROCESSING IN P200_MAINLINE
PLIIMS3 - SUCCESSFULLY ENDED
```

Updating a Segment (GHU/REPL)

In COBIMS4 we will update a record. Updating (either changing or deleting a record) always involves two steps in IMS. You must first get and lock the record you are operating on so that no other process can make updates to it. Second you issue either a REPL or DLET call.

A Get Hold Unique (GHU) call prevents any other process from making modifications to the record until you are finished with it. Similar calls are Get Hold Next (GHN) and Get Hold Next in Parent (GHNP).

For this example, let's change the promotion date on employee 9134 to Sept 1, 2016. To do that we need a GHU call with a qualified SSA that we have loaded with the employee id value of 9134.

```
MOVE '9134' TO EMP-ID-VAL
```

Here is the GHU call, and notice we are using IO-EMPLOYEE-RECORD as our segment I/O area. This is because it has the full record layout with all the fields which makes it easy to change the promotion date by field assignment.

```
CALL 'CBLTDLI' USING FOUR,
        DLI-FUNCGHU,
        PCB-MASK,
        IO-EMPLOYEE-RECORD,
        EMP-QUALIFIED-SSA
```

Once you've done the GHU call you can change the value of the promotion date.

```
MOVE '2016-09-01' TO EMPL-PRM-DTE
```

Finally you issue the REPL call. A REPL does not use any SSA since the record is already held in memory. It simply uses the segment I/O area to perform the update to the database. So you only have three parameters.

```
CALL 'CBLTDLI' USING THREE,
        DLI-FUNCREPL,
        PCB-MASK,
        IO-EMPLOYEE-RECORD
```

Here is the entire program listing for COBIMS4.

```
IDENTIFICATION DIVISION.
PROGRAM-ID. COBIMS4.
********************************************************
*     UPDATE A RECORD FROM IMS EMPLOYEE DATABASE     *
********************************************************
ENVIRONMENT DIVISION.
DATA DIVISION.
********************************************************
*  W O R K I N G   S T O R A G E   S E C T I O N     *
********************************************************
WORKING-STORAGE SECTION.
01 SEG-IO-AREA      PIC X(80).

01 IO-EMPLOYEE-RECORD.
    05  EMPL-ID-IN    PIC X(04).
    05  FILLER        PIC X(01).
    05  EMPL-LNAME    PIC X(30).
    05  FILLER        PIC X(01).
    05  EMPL-FNAME    PIC X(20).
    05  FILLER        PIC X(01).
    05  EMPL-YRS-SRV  PIC X(02).
    05  FILLER        PIC X(01).
    05  EMPL-PRM-DTE  PIC X(10).
    05  FILLER        PIC X(10).

01 DLI-FUNCTIONS.
    05 DLI-FUNCISRT  PIC X(4) VALUE 'ISRT'.
    05 DLI-FUNCGU    PIC X(4) VALUE 'GU  '.
    05 DLI-FUNCGN    PIC X(4) VALUE 'GN  '.
    05 DLI-FUNCGHU   PIC X(4) VALUE 'GHU '.
    05 DLI-FUNCGNP   PIC X(4) VALUE 'GNP '.
    05 DLI-FUNCREPL  PIC X(4) VALUE 'REPL'.
    05 DLI-FUNCDLET  PIC X(4) VALUE 'DLET'.
    05 DLI-FUNCXRST  PIC X(4) VALUE 'XRST'.
    05 DLI-FUNCCKPT  PIC X(4) VALUE 'CKPT'.

  01 EMP-UNQUALIFIED-SSA.
      05  SEGNAME    PIC X(08) VALUE 'EMPLOYEE'.
      05  FILLER     PIC X(01) VALUE ' '.

  01 EMP-QUALIFIED-SSA.
      05  SEGNAME    PIC X(08) VALUE 'EMPLOYEE'.
      05  FILLER     PIC X(01) VALUE '('.
      05  FIELD      PIC X(08) VALUE 'EMPID'.
      05  OPER       PIC X(02) VALUE ' ='.
      05  EMP-ID-VAL PIC X(04) VALUE '    '.
```

```
           05  FILLER        PIC X(01) VALUE ')'.

       01 IMS-RET-CODES.
           05 THREE           PIC S9(9) COMP VALUE +3.
           05 FOUR            PIC S9(9) COMP VALUE +4.
           05 FIVE            PIC S9(9) COMP VALUE +5.
           05 SIX             PIC S9(9) COMP VALUE +6.

       LINKAGE SECTION.
        01 PCB-MASK.
           03 DBD-NAME        PIC X(8).
           03 SEG-LEVEL       PIC XX.
           03 STATUS-CODE     PIC XX.
           03 PROC-OPT        PIC X(4).
           03 FILLER          PIC X(4).
           03 SEG-NAME        PIC X(8).
           03 KEY-FDBK        PIC S9(5) COMP.
           03 NUM-SENSEG      PIC S9(5) COMP.
           03 KEY-FDBK-AREA.
              05 EMPLOYEE-ID  PIC X(04).
              05 EMPPAYHS     PIC X(08).

       PROCEDURE DIVISION.

           INITIALIZE PCB-MASK
           ENTRY 'DLITCBL' USING PCB-MASK

           PERFORM P100-INITIALIZATION.
           PERFORM P200-MAINLINE.
           PERFORM P300-TERMINATION.
           GOBACK.

       P100-INITIALIZATION.

           DISPLAY '** PROGRAM COBIMS4 START **'
           DISPLAY 'PROCESSING IN P100-INITIALIZATION'.

       P200-MAINLINE.

           DISPLAY 'PROCESSING IN P200-MAINLINE'
           MOVE '9134' TO EMP-ID-VAL

      *    AQCUIRE THE SEGMENT WITH HOLD

           CALL 'CBLTDLI' USING FOUR,
                          DLI-FUNCGHU,
                          PCB-MASK,
                          IO-EMPLOYEE-RECORD,
```

```
                    EMP-QUALIFIED-SSA

      IF STATUS-CODE = '  '
         DISPLAY 'SUCCESSFUL GET HOLD CALL  '
         DISPLAY 'IO-EMPLOYEE-RECORD : ' IO-EMPLOYEE-RECORD

*     NOW MAKE THE CHANGE AND REPLACE THE SEGMENT

         MOVE '2016-09-01' TO EMPL-PRM-DTE

         CALL 'CBLTDLI' USING THREE,
                    DLI-FUNCREPL,
                    PCB-MASK,
                    IO-EMPLOYEE-RECORD

      IF STATUS-CODE = '  '
         DISPLAY 'SUCCESSFUL REPLACEMENT '
         DISPLAY 'IO-EMPLOYEE-RECORD : ' IO-EMPLOYEE-RECORD
      ELSE
         DISPLAY 'ERROR IN REPLACE :' STATUS-CODE
         PERFORM P400-DISPLAY-ERROR
      END-IF

   ELSE
      DISPLAY 'ERROR IN GET HOLD :' STATUS-CODE
      PERFORM P400-DISPLAY-ERROR
   END-IF.

 P300-TERMINATION.

    DISPLAY 'PROCESSING IN P300-TERMINATION'
    DISPLAY '** COBIMS4 - SUCCESSFULLY ENDED **'.

 P400-DISPLAY-ERROR.

    DISPLAY 'ERROR ENCOUNTERED - DETAIL FOLLOWS'
    DISPLAY 'SEG-IO-AREA     :' SEG-IO-AREA
    DISPLAY 'DBD-NAME1:'      DBD-NAME
    DISPLAY 'SEG-LEVEL1:'     SEG-LEVEL
    DISPLAY 'STATUS-CODE:'    STATUS-CODE
    DISPLAY 'PROC-OPT1 :'     PROC-OPT
    DISPLAY 'SEG-NAME1 :'     SEG-NAME
    DISPLAY 'KEY-FDBK1 :'     KEY-FDBK
    DISPLAY 'NUM-SENSEG1:'    NUM-SENSEG
    DISPLAY 'KEY-FDBK-AREA1:' KEY-FDBK-AREA.

*     END OF SOURCE CODE
```

Now let's compile and link, and then run the program.

```
** PROGRAM COBIMS4 START **
PROCESSING IN P100-INITIALIZATION
PROCESSING IN P200-MAINLINE
SUCCESSFUL GET HOLD CALL
IO-EMPLOYEE-RECORD : 9134 FRANKLIN            BRIANNA        00  2016-10-01
937293598
SUCCESSFUL REPLACEMENT
IO-EMPLOYEE-RECORD : 9134 FRANKLIN            BRIANNA        00 2016-09-01
937293598
PROCESSING IN P300-TERMINATION
** COBIMS4 - SUCCESSFULLY ENDED **
```

This is the basic model for doing updates to a database segment. Here is the PLI code for those of you who are following along in PLI.

```
PLIIMS4: PROCEDURE (DB_PTR_PCB) OPTIONS(MAIN);
/*******************************************************************
* PROGRAM NAME :   PLIIMS4 - UPDATE A RECORD FROM EMPLOYEE DB      *
*******************************************************************/

/*******************************************************************
/*                 W O R K I N G   S T O R A G E                   *
*******************************************************************/

    DCL ONCODE                   BUILTIN;
    DCL DB_PTR_PCB               POINTER;
    DCL PLITDLI                  EXTERNAL ENTRY;

    DCL 01 DLI_FUNCTIONS,
           05 DLI_FUNCISRT       CHAR(04) INIT ('ISRT'),
           05 DLI_FUNCGU         CHAR(04) INIT ('GU  '),
           05 DLI_FUNCGN         CHAR(04) INIT ('GN  '),
           05 DLI_FUNCGHU        CHAR(04) INIT ('GHU '),
           05 DLI_FUNCGNP        CHAR(04) INIT ('GNP '),
           05 DLI_FUNCREPL       CHAR(04) INIT ('REPL'),
           05 DLI_FUNCDLET       CHAR(04) INIT ('DLET'),
           05 DLI_FUNCXRST       CHAR(04) INIT ('XRST'),
           05 DLI_FUNCCHKP       CHAR(04) INIT ('CHKP'),
           05 DLI_FUNCROLL       CHAR(04) INIT ('ROLL');

    DCL 01 IO_EMPLOYEE_RECORD,
           05  EMPL_ID_IN        CHAR(04),
           05  FILLER1           CHAR(01),
           05  EMPL_LNAME        CHAR(30),
           05  FILLER2           CHAR(01),
           05  EMPL_FNAME        CHAR(20),
           05  FILLER3           CHAR(01),
           05  EMPL_YRS_SRV      CHAR(02),
```

```
                05  FILLER4           CHAR(01),
                05  EMPL_PRM_DTE      CHAR(10),
                05  FILLER5           CHAR(10);

        DCL 01 PCB_MASK              BASED(DB_PTR_PCB),
                05 DBD_NAME           CHAR(08),
                05 SEG_LEVEL          CHAR(02),
                05 STATUS_CODE        CHAR(02),
                05 PROC_OPT           CHAR(04),
                05 FILLER6            FIXED BIN (31),
                05 SEG_NAME           CHAR(08),
                05 KEY_FDBK           FIXED BIN (31),
                05 NUM_SENSEG         FIXED BIN (31),
                05 KEY_FDBK_AREA,
                   10 EMPLOYEE_ID     CHAR(04);

        DCL 01 EMP_UNQUALIFIED_SSA,
                05  SEGNAME           CHAR(08) INIT ('EMPLOYEE'),
                05  FILLER7           CHAR(01) INIT (' ');

        DCL 01 EMP_QUALIFIED_SSA,
                05  SEGNAME           CHAR(08) INIT('EMPLOYEE'),
                05  FILLER8           CHAR(01) INIT('('),
                05  FIELD             CHAR(08) INIT('EMPID'),
                05  OPER              CHAR(02) INIT(' ='),
                05  EMP_ID_VAL        CHAR(04) INIT('    '),
                05  FILLER9           CHAR(01) INIT(')');

        DCL THREE                    FIXED BIN (31) INIT(3);
        DCL FOUR                     FIXED BIN (31) INIT(4);
        DCL FIVE                     FIXED BIN (31) INIT(5);

/********************************************************************
/*                P R O G R A M   M A I N L I N E              *
********************************************************************/

CALL P100_INITIALIZATION;
CALL P200_MAINLINE;
CALL P300_TERMINATION;

P100_INITIALIZATION: PROC;

   PUT SKIP LIST ('PLIIMS4: UPDATE RECORD FROM EMPLOYEE DB');
   IO_EMPLOYEE_RECORD  = '';
   PCB_MASK  = '';

END P100_INITIALIZATION;


P200_MAINLINE: PROC;

   /*  SET THE EMPLOYEE SEGMENT SEARCH ARGUMENT AND CALL PLITDLI */

   EMP_ID_VAL = '9134';
```

```
          CALL PLITDLI (FOUR,
                        DLI_FUNCGHU,
                        PCB_MASK,
                        IO_EMPLOYEE_RECORD,
                        EMP_QUALIFIED_SSA) ;

      IF STATUS_CODE = '  ' THEN
         DO;
             PUT SKIP LIST ('SUCCESSFUL GET-HOLD CALL :');
             PUT SKIP DATA(IO_EMPLOYEE_RECORD);

        /*   NOW MAKE THE CHANGE AND REPLACE THE SEGMENT */

             EMPL_PRM_DTE = '2016-09-01';

             CALL PLITDLI (THREE,
                           DLI_FUNCREPL,
                           PCB_MASK,
                           IO_EMPLOYEE_RECORD) ;

             IF STATUS_CODE = '  ' THEN
                DO;
                    PUT SKIP LIST ('SUCCESSFUL REPLACE CALL :');
                    PUT SKIP DATA(IO_EMPLOYEE_RECORD);
                END;
             ELSE
                DO;
                     PUT SKIP LIST ('ERROR IN REPLACE: ' || STATUS_CODE);
                     CALL P400_DISPLAY_ERROR;
                END;
         END;

      ELSE
         DO;
             PUT SKIP LIST ('ERROR IN GET HOLD :' || STATUS_CODE);
             CALL P400_DISPLAY_ERROR;
         END;

END P200_MAINLINE;


P300_TERMINATION: PROC;

    PUT SKIP LIST ('PLIIMS4 - ENDED SUCCESSFULLY');

END P300_TERMINATION;

P400_DISPLAY_ERROR: PROC;

    PUT SKIP LIST ('ERROR ENCOUNTERED - DETAIL FOLLOWS');
    PUT SKIP LIST ('SEG_IO_AREA     :' || SEG_IO_AREA);
    PUT SKIP LIST ('DBD_NAME1:' ||  DBD_NAME);
    PUT SKIP LIST ('SEG_LEVEL1:' || SEG_LEVEL);
```

```
    PUT SKIP LIST ('STATUS_CODE:' || STATUS_CODE);
    PUT SKIP LIST ('PROC_OPT1 :' || PROC_OPT);
    PUT SKIP LIST ('SEG_NAME1 :' || SEG_NAME);
    PUT SKIP LIST ('KEY_FDBK1 :' || KEY_FDBK);
    PUT SKIP LIST ('NUM_SENSEG1:' || NUM_SENSEG);
    PUT SKIP LIST ('KEY_FDBK_AREA1:' || KEY_FDBK_AREA);

END P400_DISPLAY_ERROR;

END PLIIMS4;
```

Deleting a Segment (GHU/DLET)

For COBIMS5 we are going to delete a record. Basically the code is exactly the same as for COBIMS4 except we are deleting instead of updating a record. Let's delete employee 9134, the one we just updated. You can simply copy the COBIMS4 code and make modifications to turn it into a delete program.

Here's the source code.

```
    IDENTIFICATION DIVISION.
    PROGRAM-ID. COBIMS5.
    *****************************************************
    *     DELETE A RECORD FROM IMS EMPLOYEE DATABASE     *
    *****************************************************
    ENVIRONMENT DIVISION.
    DATA DIVISION.

    *****************************************************
    *  W O R K I N G   S T O R A G E   S E C T I O N    *
    *****************************************************
    WORKING-STORAGE SECTION.
    01 SEG-IO-AREA      PIC X(80).

    01 IO-EMPLOYEE-RECORD.
        05  EMPL-ID-IN    PIC X(04).
        05  FILLER        PIC X(01).
        05  EMPL-LNAME    PIC X(30).
        05  FILLER        PIC X(01).
        05  EMPL-FNAME    PIC X(20).
        05  FILLER        PIC X(01).
        05  EMPL-YRS-SRV  PIC X(02).
        05  FILLER        PIC X(01).
        05  EMPL-PRM-DTE  PIC X(10).
        05  FILLER        PIC X(10).

    01 DLI-FUNCTIONS.
```

```
          05 DLI-FUNCISRT  PIC X(4) VALUE 'ISRT'.
          05 DLI-FUNCGU    PIC X(4) VALUE 'GU  '.
          05 DLI-FUNCGN    PIC X(4) VALUE 'GN  '.
          05 DLI-FUNCGHU   PIC X(4) VALUE 'GHU '.
          05 DLI-FUNCGNP   PIC X(4) VALUE 'GNP '.
          05 DLI-FUNCREPL  PIC X(4) VALUE 'REPL'.
          05 DLI-FUNCDLET  PIC X(4) VALUE 'DLET'.
          05 DLI-FUNCXRST  PIC X(4) VALUE 'XRST'.
          05 DLI-FUNCCKPT  PIC X(4) VALUE 'CKPT'.

      01 EMP-UNQUALIFIED-SSA.
          05  SEGNAME      PIC X(08) VALUE 'EMPLOYEE'.
          05  FILLER       PIC X(01) VALUE ' '.

      01 EMP-QUALIFIED-SSA.
          05  SEGNAME      PIC X(08) VALUE 'EMPLOYEE'.
          05  FILLER       PIC X(01) VALUE '('.
          05  FIELD        PIC X(08) VALUE 'EMPID'.
          05  OPER         PIC X(02) VALUE ' ='.
          05  EMP-ID-VAL   PIC X(04) VALUE '    '.
          05  FILLER       PIC X(01) VALUE ')'.

      01 IMS-RET-CODES.
          05 THREE         PIC S9(9) COMP VALUE +3.
          05 FOUR          PIC S9(9) COMP VALUE +4.
          05 FIVE          PIC S9(9) COMP VALUE +5.
          05 SIX           PIC S9(9) COMP VALUE +6.

  LINKAGE SECTION.
      01 PCB-MASK.
          03 DBD-NAME      PIC X(8).
          03 SEG-LEVEL     PIC XX.
          03 STATUS-CODE   PIC XX.
          03 PROC-OPT      PIC X(4).
          03 FILLER        PIC X(4).
          03 SEG-NAME      PIC X(8).
          03 KEY-FDBK      PIC S9(5) COMP.
          03 NUM-SENSEG    PIC S9(5) COMP.
          03 KEY-FDBK-AREA.
             05 EMPLOYEE-ID  PIC X(04).
             05 EMPPAYHS     PIC X(08).

  PROCEDURE DIVISION.

      INITIALIZE PCB-MASK
      ENTRY 'DLITCBL' USING PCB-MASK

      PERFORM P100-INITIALIZATION.
```

```
        PERFORM P200-MAINLINE.
        PERFORM P300-TERMINATION.
        GOBACK.

    P100-INITIALIZATION.

        DISPLAY '** PROGRAM COBIMS5 START **'
        DISPLAY 'PROCESSING IN P100-INITIALIZATION'.

    P200-MAINLINE.

        DISPLAY 'PROCESSING IN P200-MAINLINE'
        MOVE '9134' TO EMP-ID-VAL

*       AQCUIRE THE SEGMENT WITH HOLD

        CALL 'CBLTDLI' USING FOUR,
                      DLI-FUNCGHU,
                      PCB-MASK,
                      IO-EMPLOYEE-RECORD,
                      EMP-QUALIFIED-SSA

        IF STATUS-CODE = '  '
           DISPLAY 'SUCCESSFUL GET HOLD CALL  '
           DISPLAY 'IO-EMPLOYEE-RECORD : ' IO-EMPLOYEE-RECORD

*       NOW DELETE THE SEGMENT

           CALL 'CBLTDLI' USING THREE,
                         DLI-FUNCDLET,
                         PCB-MASK,
                         IO-EMPLOYEE-RECORD

           IF STATUS-CODE = '  '
              DISPLAY 'SUCCESSFUL DELETION OF ' EMP-ID-VAL
           ELSE
              DISPLAY 'ERROR IN DELETE :' STATUS-CODE
              PERFORM P400-DISPLAY-ERROR
           END-IF

        ELSE
           DISPLAY 'ERROR IN GET HOLD :' STATUS-CODE
           PERFORM P400-DISPLAY-ERROR
        END-IF.

    P300-TERMINATION.

        DISPLAY 'PROCESSING IN P300-TERMINATION'
```

```
              DISPLAY '** COBIMS5 - SUCCESSFULLY ENDED **'.

          P400-DISPLAY-ERROR.

              DISPLAY 'ERROR ENCOUNTERED - DETAIL FOLLOWS'
              DISPLAY 'SEG-IO-AREA      :' SEG-IO-AREA
              DISPLAY 'DBD-NAME1:'       DBD-NAME
              DISPLAY 'SEG-LEVEL1:'      SEG-LEVEL
              DISPLAY 'STATUS-CODE:'     STATUS-CODE
              DISPLAY 'PROC-OPT1 :'      PROC-OPT
              DISPLAY 'SEG-NAME1 :'      SEG-NAME
              DISPLAY 'KEY-FDBK1 :'      KEY-FDBK
              DISPLAY 'NUM-SENSEG1:'     NUM-SENSEG
              DISPLAY 'KEY-FDBK-AREA1:'  KEY-FDBK-AREA.

      *      END OF SOURCE CODE
```

Now compile, link and run:

```
** PROGRAM COBIMS5 START **
PROCESSING IN P100-INITIALIZATION
PROCESSING IN P200-MAINLINE
SUCCESSFUL GET HOLD CALL
IO-EMPLOYEE-RECORD : 9134 FRANKLIN        BRIANNA              00 2016-09-01
937293598
SUCCESSFUL DELETION OF 9134
PROCESSING IN P300-TERMINATION
** COBIMS5 - SUCCESSFULLY ENDED **
```

As you can see, the record was deleted. Here is the PLI code to perform the delete.

```
PLIIMS5: PROCEDURE (DB_PTR_PCB) OPTIONS(MAIN);
/******************************************************************
* PROGRAM NAME :   PLIIMS5 - DELETE A RECORD FROM EMPLOYEE DB     *
******************************************************************/

/******************************************************************
/*               W O R K I N G   S T O R A G E                    *
******************************************************************/

   DCL ONCODE                 BUILTIN;
   DCL DB_PTR_PCB             POINTER;
   DCL PLITDLI                EXTERNAL ENTRY;

   DCL 01 DLI_FUNCTIONS,
          05 DLI_FUNCISRT     CHAR(04) INIT ('ISRT'),
          05 DLI_FUNCGU       CHAR(04) INIT ('GU  '),
          05 DLI_FUNCGN       CHAR(04) INIT ('GN  '),
          05 DLI_FUNCGHU      CHAR(04) INIT ('GHU '),
```

```
          05  DLI_FUNCGNP          CHAR(04) INIT ('GNP '),
          05  DLI_FUNCREPL         CHAR(04) INIT ('REPL'),
          05  DLI_FUNCDLET         CHAR(04) INIT ('DLET'),
          05  DLI_FUNCXRST         CHAR(04) INIT ('XRST'),
          05  DLI_FUNCCHKP         CHAR(04) INIT ('CHKP'),
          05  DLI_FUNCROLL         CHAR(04) INIT ('ROLL');

   DCL 01 IO_EMPLOYEE_RECORD,
          05  EMPL_ID              CHAR(04),
          05  FILLER1              CHAR(01),
          05  EMPL_LNAME           CHAR(30),
          05  FILLER2              CHAR(01),
          05  EMPL_FNAME           CHAR(20),
          05  FILLER3              CHAR(01),
          05  EMPL_YRS_SRV         CHAR(02),
          05  FILLER4              CHAR(01),
          05  EMPL_PRM_DTE         CHAR(10),
          05  FILLER5              CHAR(10);

   DCL 01 PCB_MASK                 BASED(DB_PTR_PCB),
          05 DBD_NAME              CHAR(08),
          05 SEG_LEVEL             CHAR(02),
          05 STATUS_CODE           CHAR(02),
          05 PROC_OPT              CHAR(04),
          05 FILLER6               FIXED BIN (31),
          05 SEG_NAME              CHAR(08),
          05 KEY_FDBK              FIXED BIN (31),
          05 NUM_SENSEG            FIXED BIN (31),
          05 KEY_FDBK_AREA,
             10 EMPLOYEE_ID        CHAR(04);

   DCL 01 EMP_UNQUALIFIED_SSA,
          05  SEGNAME              CHAR(08) INIT ('EMPLOYEE'),
          05  FILLER7              CHAR(01) INIT (' ');

   DCL 01 EMP_QUALIFIED_SSA,
          05  SEGNAME              CHAR(08) INIT('EMPLOYEE'),
          05  FILLER8              CHAR(01) INIT('('),
          05  FIELD                CHAR(08) INIT('EMPID'),
          05  OPER                 CHAR(02) INIT(' ='),
          05  EMP_ID_VAL           CHAR(04) INIT('    '),
          05  FILLER9              CHAR(01) INIT(')');

  DCL THREE                        FIXED BIN (31) INIT(3);
  DCL FOUR                         FIXED BIN (31) INIT(4);
  DCL FIVE                         FIXED BIN (31) INIT(5);
  DCL SIX                          FIXED BIN (31) INIT(6);

/*******************************************************************
/*               P R O G R A M   M A I N L I N E            *
 *******************************************************************/

CALL P100_INITIALIZATION;
CALL P200_MAINLINE;
```

69

```
        CALL P300_TERMINATION;


P100_INITIALIZATION: PROC;

    PUT SKIP LIST ('PLIIMS5: DELETE RECORD FROM EMPLOYEE DB');
    IO_EMPLOYEE_RECORD = '';
    PCB_MASK = '';

END P100_INITIALIZATION;


P200_MAINLINE: PROC;

    /*  SET THE EMPLOYEE SEGMENT SEARCH ARGUMENT AND CALL PLITDLI */

    EMP_ID_VAL = '9134';

    CALL PLITDLI (FOUR,
                  DLI_FUNCGHU,
                  PCB_MASK,
                  IO_EMPLOYEE_RECORD,
                  EMP_QUALIFIED_SSA);

    IF STATUS_CODE = '  ' THEN
       DO;
           PUT SKIP LIST ('SUCCESSFUL GET-HOLD CALL :');
           PUT SKIP DATA(IO_EMPLOYEE_RECORD);

      /*  NOW DELETE THE SEGMENT */

           CALL PLITDLI (THREE,
                         DLI_FUNCDLET,
                         PCB_MASK,
                         IO_EMPLOYEE_RECORD);

           IF STATUS_CODE = '  ' THEN
              DO;
                  PUT SKIP LIST ('SUCCESSFUL DELETE CALL :');
                  PUT SKIP LIST ('EMPLOYEE ' || EMPL_ID);
              END;
           ELSE
              DO;
                  PUT SKIP LIST ('ERROR IN DELETE: ' || STATUS_CODE);
                  CALL P400_DISPLAY_ERROR;
              END;

       END;

    ELSE
       DO;
           PUT SKIP LIST ('ERROR IN GET HOLD :' || STATUS_CODE);
           CALL P400_DISPLAY_ERROR;
       END;
```

70

```
    END P200_MAINLINE;


P300_TERMINATION: PROC;

    PUT SKIP LIST ('PLIIMS5 - ENDED SUCCESSFULLY');

END P300_TERMINATION;

P400_DISPLAY_ERROR: PROC;

    PUT SKIP LIST ('ERROR ENCOUNTERED - DETAIL FOLLOWS');
    PUT SKIP LIST ('SEG_IO_AREA       :' || SEG_IO_AREA);
    PUT SKIP LIST ('DBD_NAME1:' || DBD_NAME);
    PUT SKIP LIST ('SEG_LEVEL1:' || SEG_LEVEL);
    PUT SKIP LIST ('STATUS_CODE:' || STATUS_CODE);
    PUT SKIP LIST ('PROC_OPT1 :' || PROC_OPT);
    PUT SKIP LIST ('SEG_NAME1 :' || SEG_NAME);
    PUT SKIP LIST ('KEY_FDBK1 :' || KEY_FDBK);
    PUT SKIP LIST ('NUM_SENSEG1:' || NUM_SENSEG);
    PUT SKIP LIST ('KEY_FDBK_AREA1:' || KEY_FDBK_AREA);

END P400_DISPLAY_ERROR;

END PLIIMS5;
```

Inserting Child Segments

So far we've only dealt with root segments. That was pretty straightforward. Now let's introduce child segments. In COBIMS6 we are going to create an EMPPAY segment under each EMPLOYEE root segment. This will be similar to how we inserted root segments except we need to specify which root segment to insert the child segment under.

First, let's look at our input file:

```
    ----+----1----+----2----+----3----+----4----+----5
    ***************************** Top of Data ****
    1111     8700000  670000  362500  20170101
    1122     8200000  600000  341666  20170101
    3217     6500000  550000  270833  20170101
    4175     5500000  150000  229166  20170101
    4720     8000000  250000  333333  20170101
    4836     6200000  220000  258333  20170101
    6288     7000000  200000  291666  20170101
    7459     8500000  450000  354166  20170101
    9134     7500000  250000  312500  20170101
```

71

To decrypt here a little, the file above contains employee id numbers with annual salary, annual bonus pay, twice-per-month paycheck dollar amount, and the effective date for all this information. Let's create a record structure in COBOL for this file.

```
01  IN-EMPPAY-RECORD.
    05  EMP-ID-IN      PIC X(04).
    05  FILLER         PIC X(05).
    05  REG-PAY-IN     PIC 99999V99.
    05  FILLER         PIC X(02).
    05  BON-PAY-IN     PIC 9999V99.
    05  FILLER         PIC X(02).
    05  SEMIMTH-IN     PIC 9999V99.
    05  FILLER         PIC X(02).
    05  EFF-DATE-IN    PIC X(08).
    05  FILLER         PIC X(38).
```

We'll also need an IMS I/O area for the EMPPAY segment. How about this one? We'll map data from the input record into this I/O area before we do the ISRT action. Note that we are using packed data fields for the IMS segment. This will save some space.

```
01  IO-EMPPAY-RECORD.
    05  PAY-EFF-DATE   PIC X(8).
    05  PAY-REG-PAY    PIC S9(6)V9(2) USAGE COMP-3.
    05  PAY-BON-PAY    PIC S9(6)V9(2) USAGE COMP-3.
    05  SEMIMTH-PAY    PIC S9(6)V9(2) USAGE COMP-3.
    05  FILLER         PIC X(57).
```

Finally, we need our SSA structures. We'll be using the unqualified EMPPAY SSA, but we'll go ahead and add both the qualified and unqualified SSAs to the program.

```
01  EMPPAY-UNQUALIFIED-SSA.
    05  SEGNAME       PIC X(08) VALUE 'EMPPAY  '.
    05  FILLER        PIC X(01) VALUE ' '.

01  EMPPAY-QUALIFIED-SSA.
    05  SEGNAME       PIC X(08) VALUE 'EMPPAY  '.
    05  FILLER        PIC X(01) VALUE '('.
    05  FIELD         PIC X(08) VALUE 'EFFDATE '.
    05  OPER          PIC X(02) VALUE ' ='.
    05  EFFDATE-VAL   PIC X(08) VALUE '        '.
    05  FILLER        PIC X(01) VALUE ')'.
```

So given this information, our ISRT call should look like this. Notice that we use a qualified SSA for the EMPLOYEE root segment, and an unqualified SSA for the EMPPAY segment.

```
CALL 'CBLTDLI' USING FIVE,
       DLI-FUNCISRT,
       PCB-MASK,
       IO-EMPPAY-RECORD,
       EMP-QUALIFIED-SSA
       EMPPAY-UNQUALIFIED-SSA
```

Of course we will need a loop for reading the input pay file, and we'll need code to map the input fields to the EMPPAY fields. And we must move the employee id on the input file to the EMPLOYEE qualified SSA. Finally, recall that we deleted employee 9134, but there is a record in the input file for 9134. Have we coded to handle this missing root? We'll soon see.

Here is our completed code for COBIMS6.

```
IDENTIFICATION DIVISION.
PROGRAM-ID. COBIMS6.

*********************************************************
*    INSERT EMPLOYEE PAY RECORDS INTO THE EMPLOYEE    *
*    IMS DATABASE. ROOT KEY MUST BE SPECIFIED.        *
*********************************************************

ENVIRONMENT DIVISION.
INPUT-OUTPUT SECTION.

    FILE-CONTROL.
        SELECT EMPPAY-IN-FILE    ASSIGN TO EMPPAYFL.

DATA DIVISION.

FILE SECTION.
FD EMPPAY-IN-FILE
    RECORDING MODE IS F
    RECORD CONTAINS 80 CHARACTERS
    DATA RECORD IS IN-EMPPAY-RECORD.

    01 IN-EMPPAY-RECORD.
       05  EMP-ID-IN     PIC X(04).
       05  FILLER        PIC X(05).
       05  REG-PAY-IN    PIC 99999V99.
       05  FILLER        PIC X(02).
       05  BON-PAY-IN    PIC 9999V99.
```

```
          05  FILLER          PIC X(02).
          05  SEMIMTH-IN      PIC 9999V99.
          05  FILLER          PIC X(02).
          05  EFF-DATE-IN     PIC X(08).
          05  FILLER          PIC X(38).

******************************************************
*  W O R K I N G   S T O R A G E   S E C T I O N   *
******************************************************

WORKING-STORAGE SECTION.

01 WS-FLAGS.
   05  SW-END-OF-FILE-SWITCH   PIC X(1) VALUE 'N'.
       88  SW-END-OF-FILE                VALUE 'Y'.
       88  SW-NOT-END-OF-FILE            VALUE 'N'.

01 IO-EMPLOYEE-RECORD.
   05  EMPL-ID-IN      PIC X(04).
   05  FILLER          PIC X(01).
   05  EMPL-LNAME      PIC X(30).
   05  FILLER          PIC X(01).
   05  EMPL-FNAME      PIC X(20).
   05  FILLER          PIC X(01).
   05  EMPL-YRS-SRV    PIC X(02).
   05  FILLER          PIC X(01).
   05  EMPL-PRM-DTE    PIC X(10).
   05  FILLER          PIC X(10).

01 IO-EMPPAY-RECORD.
   05  PAY-EFF-DATE    PIC X(8).
   05  PAY-REG-PAY     PIC S9(6)V9(2) USAGE COMP-3.
   05  PAY-BON-PAY     PIC S9(6)V9(2) USAGE COMP-3.
   05  SEMIMTH-PAY     PIC S9(6)V9(2) USAGE COMP-3.
   05  FILLER          PIC X(57).

01 SEG-IO-AREA      PIC X(80).

01 DLI-FUNCTIONS.
   05 DLI-FUNCISRT PIC X(4) VALUE 'ISRT'.
   05 DLI-FUNCGU   PIC X(4) VALUE 'GU  '.
   05 DLI-FUNCGN   PIC X(4) VALUE 'GN  '.
   05 DLI-FUNCGHU  PIC X(4) VALUE 'GHU '.
   05 DLI-FUNCGNP  PIC X(4) VALUE 'GNP '.
   05 DLI-FUNCREPL PIC X(4) VALUE 'REPL'.
   05 DLI-FUNCDLET PIC X(4) VALUE 'DLET'.
   05 DLI-FUNCXRST PIC X(4) VALUE 'XRST'.
   05 DLI-FUNCCKPT PIC X(4) VALUE 'CKPT'.
```

```cobol
01 EMP-UNQUALIFIED-SSA.
   05  SEGNAME       PIC X(08) VALUE 'EMPLOYEE'.
   05  FILLER        PIC X(01) VALUE ' '.

01 EMP-QUALIFIED-SSA.
   05  SEGNAME       PIC X(08) VALUE 'EMPLOYEE'.
   05  FILLER        PIC X(01) VALUE '('.
   05  FIELD         PIC X(08) VALUE 'EMPID'.
   05  OPER          PIC X(02) VALUE ' ='.
   05  EMP-ID-VAL    PIC X(04) VALUE '    '.
   05  FILLER        PIC X(01) VALUE ')'.

01 EMPPAY-UNQUALIFIED-SSA.
   05  SEGNAME       PIC X(08) VALUE 'EMPPAY  '.
   05  FILLER        PIC X(01) VALUE ' '.

01 EMPPAY-QUALIFIED-SSA.
   05  SEGNAME       PIC X(08) VALUE 'EMPPAY  '.
   05  FILLER        PIC X(01) VALUE '('.
   05  FIELD         PIC X(08) VALUE 'EFFDATE '.
   05  OPER          PIC X(02) VALUE ' ='.
   05  EFFDATE-VAL   PIC X(08) VALUE '        '.
   05  FILLER        PIC X(01) VALUE ')'.

01 IMS-RET-CODES.
   05 THREE          PIC S9(9) COMP VALUE +3.
   05 FOUR           PIC S9(9) COMP VALUE +4.
   05 FIVE           PIC S9(9) COMP VALUE +5.
   05 SIX            PIC S9(9) COMP VALUE +6.


LINKAGE SECTION.
 01 PCB-MASK.
    03 DBD-NAME       PIC X(8).
    03 SEG-LEVEL      PIC XX.
    03 STATUS-CODE    PIC XX.
    03 PROC-OPT       PIC X(4).
    03 FILLER         PIC X(4).
    03 SEG-NAME       PIC X(8).
    03 KEY-FDBK       PIC S9(5) COMP.
    03 NUM-SENSEG     PIC S9(5) COMP.
    03 KEY-FDBK-AREA.
       05 EMPLOYEE-ID  PIC X(04).
       05 EMPPAYHS     PIC X(08).

PROCEDURE DIVISION.
```

```
        INITIALIZE PCB-MASK
        ENTRY 'DLITCBL' USING PCB-MASK

        PERFORM P100-INITIALIZATION.
        PERFORM P200-MAINLINE.
        PERFORM P300-TERMINATION.
        GOBACK.

P100-INITIALIZATION.

        DISPLAY '** PROGRAM COBIMS6 START **'
        DISPLAY 'PROCESSING IN P100-INITIALIZATION'
        OPEN INPUT EMPPAY-IN-FILE.

P200-MAINLINE.

        DISPLAY 'PROCESSING IN P200-MAINLINE'

        READ EMPPAY-IN-FILE
           AT END SET SW-END-OF-FILE TO TRUE
        END-READ

        PERFORM UNTIL SW-END-OF-FILE

           DISPLAY 'MAPPING FIELDS FOR EMPLOYEE ' EMP-ID-IN
           DISPLAY 'EFF-DATE-IN ' EFF-DATE-IN
           DISPLAY 'REG-PAY-IN  ' REG-PAY-IN
           DISPLAY 'BON-PAY-IN  ' BON-PAY-IN
           DISPLAY 'SEMIMTH-IN  ' SEMIMTH-IN
           MOVE EMP-ID-IN    TO EMP-ID-VAL
           MOVE EFF-DATE-IN  TO PAY-EFF-DATE
           MOVE REG-PAY-IN   TO PAY-REG-PAY
           MOVE BON-PAY-IN   TO PAY-BON-PAY
           MOVE SEMIMTH-IN   TO SEMIMTH-PAY

           CALL 'CBLTDLI' USING FIVE,
                DLI-FUNCISRT,
                PCB-MASK,
                IO-EMPPAY-RECORD,
                EMP-QUALIFIED-SSA
                EMPPAY-UNQUALIFIED-SSA

           IF STATUS-CODE = ' '
              DISPLAY 'SUCCESSFUL INSERT-REC FOR EMP: ' EMP-ID-VAL
              DISPLAY 'SUCCESSFUL INSERT-REC VALUES : '
              IO-EMPPAY-RECORD
           ELSE
              PERFORM P400-DISPLAY-ERROR
```

```
          END-IF

              READ EMPPAY-IN-FILE
                 AT END SET SW-END-OF-FILE TO TRUE
              END-READ

          END-PERFORM.

      P300-TERMINATION.

          DISPLAY 'PROCESSING IN P300-TERMINATION'
          CLOSE EMPPAY-IN-FILE.
          DISPLAY '** COBIMS6 - SUCCESSFULLY ENDED **'.

      P400-DISPLAY-ERROR.

          DISPLAY 'ERROR ENCOUNTERED - DETAIL FOLLOWS'
          DISPLAY 'DBD-NAME1:'     DBD-NAME
          DISPLAY 'SEG-LEVEL1:'    SEG-LEVEL
          DISPLAY 'STATUS-CODE:'   STATUS-CODE
          DISPLAY 'PROC-OPT1 :'    PROC-OPT
          DISPLAY 'SEG-NAME1 :'    SEG-NAME
          DISPLAY 'KEY-FDBK1 :'    KEY-FDBK
          DISPLAY 'NUM-SENSEG1:'   NUM-SENSEG
          DISPLAY 'KEY-FDBK-AREA1:' KEY-FDBK-AREA.

      *    END OF SOURCE CODE
```

Compile and link, and then run the program. Here is the output.

```
** PROGRAM COBIMS6 START **
PROCESSING IN P100-INITIALIZATION
PROCESSING IN P200-MAINLINE
MAPPING FIELDS FOR EMPLOYEE 1111
EFF-DATE-IN 20170101
REG-PAY-IN  8700000
BON-PAY-IN  670000
SEMIMTH-IN  362500
SUCCESSFUL INSERT-REC FOR EMP: 1111
SUCCESSFUL INSERT-REC VALUES : 20170101 g              &
MAPPING FIELDS FOR EMPLOYEE 1122
EFF-DATE-IN 20170101
REG-PAY-IN  8200000
BON-PAY-IN  600000
SEMIMTH-IN  341666
SUCCESSFUL INSERT-REC FOR EMP: 1122
SUCCESSFUL INSERT-REC VALUES : 20170101 b                %
```

```
MAPPING FIELDS FOR EMPLOYEE 3217
EFF-DATE-IN 20170101
REG-PAY-IN  6500000
BON-PAY-IN  550000
SEMIMTH-IN  270833
SUCCESSFUL INSERT-REC FOR EMP: 3217
SUCCESSFUL INSERT-REC VALUES : 20170101          &      c
MAPPING FIELDS FOR EMPLOYEE 4175
EFF-DATE-IN 20170101
REG-PAY-IN  5500000
BON-PAY-IN  150000
SEMIMTH-IN  229166
SUCCESSFUL INSERT-REC FOR EMP: 4175
SUCCESSFUL INSERT-REC VALUES : 20170101          &       %
MAPPING FIELDS FOR EMPLOYEE 4720
EFF-DATE-IN 20170101
REG-PAY-IN  8000000
BON-PAY-IN  250000
SEMIMTH-IN  333333
SUCCESSFUL INSERT-REC FOR EMP: 4720
SUCCESSFUL INSERT-REC VALUES : 20170101          &
MAPPING FIELDS FOR EMPLOYEE 4836
EFF-DATE-IN 20170101
REG-PAY-IN  6200000
BON-PAY-IN  220000
SEMIMTH-IN  258333
SUCCESSFUL INSERT-REC FOR EMP: 4836
SUCCESSFUL INSERT-REC VALUES : 20170101
MAPPING FIELDS FOR EMPLOYEE 6288
EFF-DATE-IN 20170101
REG-PAY-IN  7000000
BON-PAY-IN  200000
SEMIMTH-IN  291666
SUCCESSFUL INSERT-REC FOR EMP: 6288
SUCCESSFUL INSERT-REC VALUES : 20170101               j %
MAPPING FIELDS FOR EMPLOYEE 7459
EFF-DATE-IN 20170101
REG-PAY-IN  8500000
BON-PAY-IN  450000
SEMIMTH-IN  354166
SUCCESSFUL INSERT-REC FOR EMP: 7459
SUCCESSFUL INSERT-REC VALUES : 20170101 e      &       %
MAPPING FIELDS FOR EMPLOYEE 9134
EFF-DATE-IN 20170101
REG-PAY-IN  7500000
BON-PAY-IN  250000
SEMIMTH-IN  312500
```
ERROR ENCOUNTERED - DETAIL FOLLOWS

```
DBD-NAME1:EMPLOYEE
SEG-LEVEL1:00
STATUS-CODE:GE
PROC-OPT1 :AP
SEG-NAME1 :
KEY-FDBK1 :00000
NUM-SENSEG1:00004
KEY-FDBK-AREA1:     20170101
PROCESSING IN P300-TERMINATION
** COBIMS6 - SUCCESSFULLY ENDED **
```

The bolded text above shows that our error code caught the missing root segment and reported it. In this case we took a "soft landing" by not terminating the program. In the real world we might have forced an abend.[3] Or we might possibly have written the record to an exception report for someone to review and correct.

So now you have a model for inserting new data to child segments in an IMS database. Here is the PLI code that corresponds to COBIMS6.

```
PLIIMS6: PROCEDURE (DB_PTR_PCB) OPTIONS(MAIN);
/*******************************************************************
* PROGRAM NAME :  PLIIMS6 - INSERT EMPLOYEE PAY RECORDS INTO THE  *
*                 EMPLOYEE IMS DB. ROOT KEY MUST BE SPECIFIED.    *
*******************************************************************/
/*******************************************************************
/*              F I L E S   U S E D                               *
*******************************************************************/

   DCL EMPPAYFL FILE RECORD SEQL INPUT;

/*******************************************************************
/*              W O R K I N G   S T O R A G E                     *
*******************************************************************/

   DCL SW_END_OF_FILE          STATIC BIT(01) INIT('0'B);
   DCL ONCODE                  BUILTIN;
   DCL DB_PTR_PCB              POINTER;


   DCL PLITDLI                 EXTERNAL ENTRY;

   DCL 01 DLI_FUNCTIONS,
          05 DLI_FUNCISRT      CHAR(04) INIT ('ISRT'),
          05 DLI_FUNCGU        CHAR(04) INIT ('GU  '),
```

---

3 You can force an abend with a memory dump by calling LE program CEE3DMP. Details for how to do that are at the link below. We will only take soft abends in this text, so we won't abend with CEE3DMP.

https://www.ibm.com/support/knowledgecenter/en/SSLTBW_2.3.0/com.ibm.zos.v2r3.ceea100/ceea1mst78.htm

```
            05  DLI_FUNCGN          CHAR(04) INIT ('GN  '),
            05  DLI_FUNCGHU         CHAR(04) INIT ('GHU '),
            05  DLI_FUNCGNP         CHAR(04) INIT ('GNP '),
            05  DLI_FUNCREPL        CHAR(04) INIT ('REPL'),
            05  DLI_FUNCDLET        CHAR(04) INIT ('DLET'),
            05  DLI_FUNCXRST        CHAR(04) INIT ('XRST'),
            05  DLI_FUNCCHKP        CHAR(04) INIT ('CHKP'),
            05  DLI_FUNCROLL        CHAR(04) INIT ('ROLL');


    DCL 01 IN_EMPPAY_RECORD,
            05  EMP_ID_IN           CHAR(04),
            05  FILLER1             CHAR(05),
            05  REG_PAY_IN          PIC '99999V99',
            05  FILLER2             CHAR(02),
            05  BON_PAY_IN          PIC '9999V99',
            05  FILLER3             CHAR(02),
            05  SEMIMTH_IN          PIC '9999V99',
            05  FILLER4             CHAR(02),
            05  EFF_DATE_IN         CHAR(08),
            05  FILLER5             CHAR(38);

    DCL 01 IO_EMPPAY_RECORD,
            05  PAY_EFF_DATE        CHAR(8),
            05  PAY_REG_PAY         FIXED DEC (8,2),
            05  PAY_BON_PAY         FIXED DEC (8,2),
            05  SEMIMTH_PAY         FIXED DEC (8,2),
            05  FILLER55            CHAR(57);

    DCL 01 PCB_MASK                 BASED(DB_PTR_PCB),
            05  DBD_NAME            CHAR(08),
            05  SEG_LEVEL           CHAR(02),
            05  STATUS_CODE         CHAR(02),
            05  PROC_OPT            CHAR(04),
            05  FILLER6             FIXED BIN (31),
            05  SEG_NAME            CHAR(08),
            05  KEY_FDBK            FIXED BIN (31),
            05  NUM_SENSEG          FIXED BIN (31),
            05  KEY_FDBK_AREA,
            10  EMPLOYEE_ID     CHAR(04),
            10  EMP_PAY_DATE    CHAR(08);

    DCL 01 EMP_UNQUALIFIED_SSA,
            05  SEGNAME             CHAR(08) INIT ('EMPLOYEE'),
            05  FILLER7             CHAR(01) INIT (' ');

    DCL 01 EMP_QUALIFIED_SSA,
            05  SEGNAME             CHAR(08) INIT('EMPLOYEE'),
            05  FILLER8             CHAR(01) INIT('('),
            05  FIELD               CHAR(08) INIT('EMPID'),
            05  OPER                CHAR(02) INIT(' ='),
            05  EMP_ID_VAL          CHAR(04) INIT('    '),
            05  FILLER9             CHAR(01) INIT(')');
```

80

```
   DCL 01 EMPPAY_UNQUALIFIED_SSA,
           05   SEGNAME              CHAR(08) INIT('EMPPAY  '),
           05   FILLER10             CHAR(01) INIT(' ');

   DCL 01 EMPPAY_QUALIFIED_SSA,
           05   SEGNAME              CHAR(08) INIT('EMPPAY  '),
           05   FILLER11             CHAR(01) INIT('('),
           05   FIELD                CHAR(08) INIT('EFFDATE '),
           05   OPER                 CHAR(02) INIT(' ='),
           05   EFFDATE_VAL          CHAR(08) INIT('        '),
           05   FILLER12             CHAR(01) INIT(')');

   DCL SEG_IO_AREA                   CHAR(80) INIT (' ');

   DCL THREE                         FIXED BIN (31) INIT(3);
   DCL FOUR                          FIXED BIN (31) INIT(4);
   DCL FIVE                          FIXED BIN (31) INIT(5);
   DCL SIX                           FIXED BIN (31) INIT(6);

/**********************************************************************
/*                 O N   C O N D I T I O N S                         *
**********************************************************************/

   ON ENDFILE (EMPIFILE) SW_END_OF_FILE = '1'B;

/**********************************************************************
/*                 P R O G R A M   M A I N L I N E                   *
**********************************************************************/

CALL P100_INITIALIZATION;
CALL P200_MAINLINE;
CALL P300_TERMINATION;

P100_INITIALIZATION: PROC;

    PUT SKIP LIST ('PLIIMS6: INSERT RECORDS');
    OPEN FILE (EMPPAYFL);

    IN_EMPPAY_RECORD  = '';
    PCB_MASK = '';

END P100_INITIALIZATION;

P200_MAINLINE: PROC;

    /*  MAIN LOOP - READ THE INPUT FILE, LOAD THE OUTPUT
                    STRUCTURE AND WRITE PAY RECORD TO OUTPUT */

    READ FILE (EMPPAYFL) INTO (IN_EMPPAY_RECORD);

    DO WHILE (¬SW_END_OF_FILE);

        /* ASSIGN KEY FOR EMPLOYEE LEVEL SSA */
```

```
            EMP_ID_VAL    = EMP_ID_IN;

        /* ASSIGN PAY FIELDS */

        PAY_EFF_DATE = EFF_DATE_IN;
        PAY_REG_PAY  = REG_PAY_IN;
        PAY_BON_PAY  = BON_PAY_IN;
        SEMIMTH_PAY  = SEMIMTH_IN;

        CALL PLITDLI (FIVE,
                      DLI_FUNCISRT,
                      PCB_MASK,
                      IO_EMPPAY_RECORD,
                      EMP_QUALIFIED_SSA,
                      EMPPAY_UNQUALIFIED_SSA);

        IF STATUS_CODE = '  ' THEN
           DO;
              PUT SKIP LIST ('SUCCESSFUL INSERT PAY REC:');
              PUT SKIP DATA (IO_EMPPAY_RECORD);
           END;
        ELSE
           DO;
              CALL P400_DISPLAY_ERROR;
              RETURN;
           END;

        READ FILE (EMPPAYFL) INTO (IN_EMPPAY_RECORD);

    END; /* DO WHILE */

END P200_MAINLINE;

P300_TERMINATION: PROC;

    CLOSE FILE(EMPPAYFL);

    PUT SKIP LIST ('PLIIMS6 - SUCCESSFULLY ENDED');

END P300_TERMINATION;

P400_DISPLAY_ERROR: PROC;

    PUT SKIP LIST ('ERROR ENCOUNTERED - DETAIL FOLLOWS');
    PUT SKIP DATA (IO_EMPPAY_RECORD);
    PUT SKIP LIST ('DBD_NAME1:' ||  DBD_NAME);
    PUT SKIP LIST ('SEG_LEVEL1:' || SEG_LEVEL);
    PUT SKIP LIST ('STATUS_CODE:' || STATUS_CODE);
    PUT SKIP LIST ('PROC_OPT1 :' || PROC_OPT);
    PUT SKIP LIST ('SEG_NAME1 :' || SEG_NAME);
    PUT SKIP LIST ('KEY_FDBK1 :' || KEY_FDBK);
    PUT SKIP LIST ('NUM_SENSEG1:' || NUM_SENSEG);
    PUT SKIP LIST ('KEY_FDBK_AREA1:' || KEY_FDBK_AREA);
```

```
END P400_DISPLAY_ERROR;

END PLIIMS6;
```

Reading Child Segments Sequentially (GNP)

Now let's use COBIMS7 to read back the records we just added to the database. We can traverse the database using GN for the root segments and GNP (Get Next Within Parent) calls for the children. So we'll borrow the code from COBIMS3 for walking through the root segments. And then we'll add code for retrieving GNP.

Keep in mind that we've only added a single EMPPAY child under each root segment. If there were more than one child, our code would need to allow for that. But for now, our spec will ask us to simply get a root and then get the first child under that root. Then we will display the pay information for the employee.

We already know how to traverse the root segment. So once we get a root segment, we need to take the EMP-ID returned in the IO-EMPLOYEE-RECORD and use it to set the qualified SSA for EMPLOYEE. We could use the unqualified SSA for the EMPPAY segment, but since we already know the exact key we can as easily use the qualified SSA. And we'll load the segment data into the IO-EMPPAY-RECORD I/O area. This is what our call will look like.

```
        MOVE EMPL-ID-IN TO EMP-ID-VAL
        MOVE '20170101' TO EFFDATE-VAL

        CALL 'CBLTDLI' USING FIVE,
             DLI-FUNCGNP,
             PCB-MASK,
             IO-EMPPAY-RECORD,
             EMP-QUALIFIED-SSA,
             EMPPAY-UNQUALIFIED-SSA
```

Other than that, our program doesn't need to use any new techniques. Here is the completed program listing.

```
        IDENTIFICATION DIVISION.
        PROGRAM-ID. COBIMS7.
        *****************************************************
        *   WALK THROUGH THE EMPLOYEE AND EMPPAY SEGS OF    *
        *   THE ENTIRE EMPLOYEE IMS DATABASE.               *
```

```
*********************************************************
 ENVIRONMENT DIVISION.
 INPUT-OUTPUT SECTION.
 DATA DIVISION.

*********************************************************
*    W O R K I N G    S T O R A G E    S E C T I O N    *
*********************************************************

 WORKING-STORAGE SECTION.

   01 WS-FLAGS.
      05  SW-END-OF-DB-SWITCH      PIC X(1) VALUE 'N'.
          88  SW-END-OF-DB                  VALUE 'Y'.
          88  SW-NOT-END-OF-DB              VALUE 'N'.

   01 IO-EMPLOYEE-RECORD.
      05  EMPL-ID-IN    PIC X(04).
      05  FILLER        PIC X(01).
      05  EMPL-LNAME    PIC X(30).
      05  FILLER        PIC X(01).
      05  EMPL-FNAME    PIC X(20).
      05  FILLER        PIC X(01).
      05  EMPL-YRS-SRV  PIC X(02).
      05  FILLER        PIC X(01).
      05  EMPL-PRM-DTE  PIC X(10).
      05  FILLER        PIC X(10).

   01 IO-EMPPAY-RECORD.
      05  PAY-EFF-DATE  PIC X(8).
      05  PAY-REG-PAY   PIC S9(6)V9(2) USAGE COMP-3.
      05  PAY-BON-PAY   PIC S9(6)V9(2) USAGE COMP-3.
      05  SEMIMTH-PAY   PIC S9(6)V9(2) USAGE COMP-3.
      05  FILLER        PIC X(57).

   01 DISPLAY-EMPLOYEE-PIC.
      05  DIS-REG-PAY   PIC ZZ999.99-.
      05  DIS-BON-PAY   PIC ZZ999.99-.
      05  DIS-SMT-PAY   PIC ZZ999.99-.

   01 EMP-UNQUALIFIED-SSA.
      05  SEGNAME    PIC X(08) VALUE 'EMPLOYEE'.
      05  FILLER     PIC X(01) VALUE ' '.

   01 EMP-QUALIFIED-SSA.
      05  SEGNAME    PIC X(08) VALUE 'EMPLOYEE'.
      05  FILLER     PIC X(01) VALUE '('.
      05  FIELD      PIC X(08) VALUE 'EMPID'.
```

```
         05  OPER        PIC X(02) VALUE ' ='.
         05  EMP-ID-VAL  PIC X(04) VALUE '    '.
         05  FILLER      PIC X(01) VALUE ')'.

     01 EMPPAY-UNQUALIFIED-SSA.
         05  SEGNAME     PIC X(08) VALUE 'EMPPAY  '.
         05  FILLER      PIC X(01) VALUE ' '.

     01 EMPPAY-QUALIFIED-SSA.
         05  SEGNAME     PIC X(08) VALUE 'EMPPAY  '.
         05  FILLER      PIC X(01) VALUE '('.
         05  FIELD       PIC X(08) VALUE 'EFFDATE '.
         05  OPER        PIC X(02) VALUE ' ='.
         05  EFFDATE-VAL PIC X(08) VALUE '        '.
         05  FILLER      PIC X(01) VALUE ')'.

     01 DLI-FUNCTIONS.
         05 DLI-FUNCISRT PIC X(4) VALUE 'ISRT'.
         05 DLI-FUNCGU   PIC X(4) VALUE 'GU  '.
         05 DLI-FUNCGN   PIC X(4) VALUE 'GN  '.
         05 DLI-FUNCGHU  PIC X(4) VALUE 'GHU '.
         05 DLI-FUNCGNP  PIC X(4) VALUE 'GNP '.
         05 DLI-FUNCREPL PIC X(4) VALUE 'REPL'.
         05 DLI-FUNCDLET PIC X(4) VALUE 'DLET'.
         05 DLI-FUNCXRST PIC X(4) VALUE 'XRST'.
         05 DLI-FUNCCKPT PIC X(4) VALUE 'CKPT'.

      01 IMS-RET-CODES.
         05 THREE         PIC S9(9) COMP VALUE +3.
         05 FOUR          PIC S9(9) COMP VALUE +4.
         05 FIVE          PIC S9(9) COMP VALUE +5.
         05 SIX           PIC S9(9) COMP VALUE +6.


     LINKAGE SECTION.
      01 PCB-MASK.
         03 DBD-NAME      PIC X(8).
         03 SEG-LEVEL     PIC XX.
         03 STATUS-CODE   PIC XX.
         03 PROC-OPT      PIC X(4).
         03 FILLER        PIC X(4).
         03 SEG-NAME      PIC X(8).
         03 KEY-FDBK      PIC S9(5) COMP.
         03 NUM-SENSEG    PIC S9(5) COMP.
         03 KEY-FDBK-AREA.
            05 EMPLOYEE-KEY PIC X(04).
            05 EMPPAYHS-KEY PIC X(08).
```

```
    PROCEDURE DIVISION.

        INITIALIZE PCB-MASK
        ENTRY 'DLITCBL' USING PCB-MASK

        PERFORM P100-INITIALIZATION.
        PERFORM P200-MAINLINE.
        PERFORM P300-TERMINATION.
        GOBACK.

    P100-INITIALIZATION.

        DISPLAY '** PROGRAM COBIMS7 START **'
        DISPLAY 'PROCESSING IN P100-INITIALIZATION'.

*       DO INITIAL DB READ FOR FIRST EMPLOYEE RECORD

        CALL 'CBLTDLI' USING FOUR,
            DLI-FUNCGN,
            PCB-MASK,
            IO-EMPLOYEE-RECORD,
            EMP-UNQUALIFIED-SSA

        IF STATUS-CODE = '  ' THEN
           NEXT SENTENCE
        ELSE
           IF STATUS-CODE = 'GB' THEN
              SET SW-END-OF-DB TO TRUE
              DISPLAY 'END OF DATABASE :'
           ELSE
              PERFORM P400-DISPLAY-ERROR
              GOBACK
           END-IF

        END-IF.

    P200-MAINLINE.

        DISPLAY 'PROCESSING IN P200-MAINLINE'

*       CHECK STATUS CODE AND FIRST RECORD

        IF SW-END-OF-DB THEN
           DISPLAY 'NO RECORDS TO PROCESS!!'
        ELSE
           DISPLAY 'SUCCESSFUL READ :' IO-EMPLOYEE-RECORD
           PERFORM UNTIL SW-END-OF-DB
              PERFORM P500-GET-PAY-SEG
```

86

```
            CALL 'CBLTDLI' USING FOUR,
                  DLI-FUNCGN,
                  PCB-MASK,
                  IO-EMPLOYEE-RECORD,
                  EMP-UNQUALIFIED-SSA

            IF STATUS-CODE = 'GB' THEN
               SET SW-END-OF-DB TO TRUE
               DISPLAY 'END OF DATABASE'
            ELSE
               IF STATUS-CODE NOT EQUAL SPACES THEN
                  PERFORM P400-DISPLAY-ERROR
               ELSE
                  DISPLAY 'SUCCESSFUL READ :' IO-EMPLOYEE-RECORD
               END-IF
            END-IF

        END-PERFORM.

    DISPLAY 'FINISHED PROCESSING IN P200-MAINLINE'.

P300-TERMINATION.

    DISPLAY 'PROCESSING IN P300-TERMINATION'
    DISPLAY '** COBIMS7 - SUCCESSFULLY ENDED **'.

P400-DISPLAY-ERROR.

    DISPLAY 'PROCESSING IN P400-DISPLAY-ERROR'
    DISPLAY 'ERROR ENCOUNTERED - DETAIL FOLLOWS'
    DISPLAY 'DBD-NAME1:'      DBD-NAME
    DISPLAY 'SEG-LEVEL1:'     SEG-LEVEL
    DISPLAY 'STATUS-CODE:'    STATUS-CODE
    DISPLAY 'PROC-OPT1 :'     PROC-OPT
    DISPLAY 'SEG-NAME1 :'     SEG-NAME
    DISPLAY 'KEY-FDBK1 :'     KEY-FDBK
    DISPLAY 'NUM-SENSEG1:'    NUM-SENSEG
    DISPLAY 'KEY-FDBK-AREA1:' KEY-FDBK-AREA.

P500-GET-PAY-SEG.

    DISPLAY 'PROCESSING IN P500-GET-PAY-SEG'

    MOVE EMPL-ID-IN TO EMP-ID-VAL
    MOVE '20170101' TO EFFDATE-VAL

    CALL 'CBLTDLI' USING FIVE,
```

```
                    DLI-FUNCGNP,
                    PCB-MASK,
                    IO-EMPPAY-RECORD,
                    EMP-QUALIFIED-SSA,
                    EMPPAY-QUALIFIED-SSA

            IF STATUS-CODE NOT EQUAL SPACES THEN
               PERFORM P400-DISPLAY-ERROR
            ELSE
*           MAP FIELDS
               MOVE PAY-REG-PAY TO DIS-REG-PAY
               MOVE PAY-BON-PAY TO DIS-BON-PAY
               MOVE SEMIMTH-PAY TO DIS-SMT-PAY
               DISPLAY 'SUCCESSFUL PAY READ :'
               DISPLAY '   EFFECTIVE DATE = ' PAY-EFF-DATE
               DISPLAY '   PAY-REG-PAY = ' DIS-REG-PAY
               DISPLAY '   PAY-BON-PAY = ' DIS-BON-PAY
               DISPLAY '   SEMIMTH-PAY = ' DIS-SMT-PAY
            END-IF.

*        END OF SOURCE CODE
```

Once again, let's compile and link, and then run the program. Here is the output showing both root and child segments.

```
** PROGRAM COBIMS7 START **
PROCESSING IN P100-INITIALIZATION
PROCESSING IN P200-MAINLINE
SUCCESSFUL READ :1111 VEREEN                    CHARLES              12 201
PROCESSING IN P500-GET-PAY-SEG
SUCCESSFUL PAY READ :
   EFFECTIVE DATE = 20170101
   PAY-REG-PAY = 87000.00
   PAY-BON-PAY =  6700.00
   SEMIMTH-PAY =  3625.00
SUCCESSFUL READ :1122 JENKINS                   DEBORAH              05 201
PROCESSING IN P500-GET-PAY-SEG
SUCCESSFUL PAY READ :
   EFFECTIVE DATE = 20170101
   PAY-REG-PAY = 82000.00
   PAY-BON-PAY =  6000.00
   SEMIMTH-PAY =  3416.66
SUCCESSFUL READ :3217 JOHNSON                   EDWARD               04 201
PROCESSING IN P500-GET-PAY-SEG
SUCCESSFUL PAY READ :
   EFFECTIVE DATE = 20170101
   PAY-REG-PAY = 65000.00
   PAY-BON-PAY =  5500.00
   SEMIMTH-PAY =  2708.33
SUCCESSFUL READ :4175 TURNBULL                  FRED                 01 201
```

```
PROCESSING IN P500-GET-PAY-SEG
SUCCESSFUL PAY READ :
   EFFECTIVE DATE = 20170101
   PAY-REG-PAY = 55000.00
   PAY-BON-PAY =  1500.00
   SEMIMTH-PAY =  2291.66
SUCCESSFUL READ :4720 SCHULTZ                    TIM              09 201
PROCESSING IN P500-GET-PAY-SEG
SUCCESSFUL PAY READ :
   EFFECTIVE DATE = 20170101
   PAY-REG-PAY = 80000.00
   PAY-BON-PAY =  2500.00
   SEMIMTH-PAY =  3333.33
SUCCESSFUL READ :4836 SMITH                      SANDRA           03 201
PROCESSING IN P500-GET-PAY-SEG
SUCCESSFUL PAY READ :
   EFFECTIVE DATE = 20170101
   PAY-REG-PAY = 62000.00
   PAY-BON-PAY =  2200.00
   SEMIMTH-PAY =  2583.33
SUCCESSFUL READ :6288 WILLARD                    JOE              06 201
PROCESSING IN P500-GET-PAY-SEG
SUCCESSFUL PAY READ :
   EFFECTIVE DATE = 20170101
   PAY-REG-PAY = 70000.00
   PAY-BON-PAY =  2000.00
   SEMIMTH-PAY =  2916.66
SUCCESSFUL READ :7459 STEWART                    BETTY            07 201
PROCESSING IN P500-GET-PAY-SEG
SUCCESSFUL PAY READ :
   EFFECTIVE DATE = 20170101
   PAY-REG-PAY = 85000.00
   PAY-BON-PAY =  4500.00
   SEMIMTH-PAY =  3541.66
END OF DATABASE
FINISHED PROCESSING IN P200-MAINLINE
PROCESSING IN P300-TERMINATION
** COBIMS7 - SUCCESSFULLY ENDED **
```

Here is the PLI source code for this program:

```
PLIIMS7: PROCEDURE (DB_PTR_PCB) OPTIONS(MAIN);
/********************************************************************
* PROGRAM NAME: PLIIMS7 - WALK THROUGH THE EMPLOYEE AND EMPPAY      *
*                         SEGMENTS OF THE EMPLOYEE IMS DATABASE.    *
********************************************************************/


/********************************************************************
/*                 W O R K I N G   S T O R A G E                    *
********************************************************************/

   DCL SW_END_OF_DB            STATIC BIT(01) INIT('0'B);
   DCL SW_NO_MORE_SEGS         STATIC BIT(01) INIT('0'B);
```

```
DCL ONCODE                        BUILTIN;
DCL DB_PTR_PCB                    POINTER;

DCL PLITDLI                       EXTERNAL ENTRY;

DCL 01 DLI_FUNCTIONS,
        05 DLI_FUNCISRT           CHAR(04) INIT ('ISRT'),
        05 DLI_FUNCGU             CHAR(04) INIT ('GU  '),
        05 DLI_FUNCGN             CHAR(04) INIT ('GN  '),
        05 DLI_FUNCGHU            CHAR(04) INIT ('GHU '),
        05 DLI_FUNCGNP            CHAR(04) INIT ('GNP '),
        05 DLI_FUNCREPL           CHAR(04) INIT ('REPL'),
        05 DLI_FUNCDLET           CHAR(04) INIT ('DLET'),
        05 DLI_FUNCXRST           CHAR(04) INIT ('XRST'),
        05 DLI_FUNCCHKP           CHAR(04) INIT ('CHKP'),
        05 DLI_FUNCROLL           CHAR(04) INIT ('ROLL');

DCL 01 IO_EMPLOYEE_RECORD,
        05  EMPL_ID_IN            CHAR(04),
        05  FILLER1               CHAR(01),
        05  EMPL_LNAME            CHAR(30),
        05  FILLER2               CHAR(01),
        05  EMPL_FNAME            CHAR(20),
        05  FILLER3               CHAR(01),
        05  EMPL_YRS_SRV          CHAR(02),
        05  FILLER4               CHAR(01),
        05  EMPL_PRM_DTE          CHAR(10),
        05  FILLER5               CHAR(10);

DCL 01 IO_EMPPAY_RECORD,
        05  PAY_EFF_DATE          CHAR(8),
        05  PAY_REG_PAY           FIXED DEC (8,2),
        05  PAY_BON_PAY           FIXED DEC (8,2),
        05  SEMIMTH_PAY           FIXED DEC (8,2),
        05  FILLER6               CHAR(57);

DCL 01 IO_EMPPAYHS_RECORD.
        05  PAY-DATE              CHAR(8),
        05  PAY-ANN-PAY           FIXED DEC (8,2),
        05  PAY-AMT               FIXED DEC (8,2),
        05  FILLER65              CHAR(62);

DCL 01 PCB_MASK                   BASED(DB_PTR_PCB),
        05 DBD_NAME               CHAR(08),
        05 SEG_LEVEL              CHAR(02),
        05 STATUS_CODE            CHAR(02),
        05 PROC_OPT               CHAR(04),
        05 FILLER99               FIXED BIN (31),
        05 SEG_NAME               CHAR(08),
        05 KEY_FDBK               FIXED BIN (31),
        05 NUM_SENSEG             FIXED BIN (31),
        05 KEY_FDBK_AREA,
            10 EMPLOYEE_ID        CHAR(04);
```

```
        DCL 01 EMP_UNQUALIFIED_SSA,
               05   SEGNAME            CHAR(08) INIT ('EMPLOYEE'),
               05   FILLER7            CHAR(01) INIT (' ');

        DCL 01 EMP_QUALIFIED_SSA,
               05   SEGNAME            CHAR(08) INIT('EMPLOYEE'),
               05   FILLER8            CHAR(01) INIT('('),
               05   FIELD              CHAR(08) INIT('EMPID'),
               05   OPER               CHAR(02) INIT(' ='),
               05   EMP_ID_VAL         CHAR(04) INIT('    '),
               05   FILLER9            CHAR(01) INIT(')');

        DCL 01 EMPPAY_UNQUALIFIED_SSA,
               05   SEGNAME            CHAR(08) INIT('EMPPAY  '),
               05   FILLER10           CHAR(01) INIT(' ');

        DCL 01 EMPPAY_QUALIFIED_SSA,
               05   SEGNAME            CHAR(08) INIT('EMPPAY  '),
               05   FILLER11           CHAR(01) INIT('('),
               05   FIELD              CHAR(08) INIT('EFFDATE '),
               05   OPER               CHAR(02) INIT(' ='),
               05   EFFDATE_VAL        CHAR(08) INIT('        '),
               05   FILLER12           CHAR(01) INIT(')');

        DCL THREE                      FIXED BIN (31) INIT(3);
        DCL FOUR                       FIXED BIN (31) INIT(4);
        DCL FIVE                       FIXED BIN (31) INIT(5);
        DCL SIX                        FIXED BIN (31) INIT(6);

/**********************************************************************
/*              P R O G R A M   M A I N L I N E              *
**********************************************************************/

CALL P100_INITIALIZATION;
CALL P200_MAINLINE;
CALL P300_TERMINATION;

P100_INITIALIZATION: PROC;

    PUT SKIP LIST ('PLIIMS7: TRAVERSE EMPLOYEE DATABASE PAY SEGS');
    PUT SKIP LIST ('PROCESSING IN P100-INITIALIZATION');

    PCB_MASK = '';
    IO_EMPLOYEE_RECORD  = '';
    IO_EMPPAY_RECORD  = '';

 /* DO INITIAL DB READ FOR FIRST EMPLOYEE RECORD */

    CALL PLITDLI (FOUR,
                  DLI_FUNCGN,
                  PCB_MASK,
                  IO_EMPLOYEE_RECORD,
                  EMP_UNQUALIFIED_SSA);
```

```
      IF STATUS_CODE = '  ' THEN;
      ELSE
         IF STATUS_CODE = 'GB' THEN
            DO;
                SW_END_OF_DB = '1'B;
                PUT SKIP LIST ('** END OF DATABASE');
            END;
         ELSE
            DO;
                CALL P400_DISPLAY_ERROR;
                RETURN;
            END;

END P100_INITIALIZATION;

P200_MAINLINE: PROC;

    /*  MAIN LOOP - CYCLE THROUGH ALL ROOT SEGMENTS IN THE DB,
                    DISPLAYING THE DATA RETRIEVED              */

        IF SW_END_OF_DB THEN
            PUT SKIP LIST ('NO RECORDS TO PROCESS!!');
        ELSE
            DO WHILE (¬SW_END_OF_DB);
                PUT SKIP LIST ('SUCCESSFUL EMPLOYEE READ : ' ||
                EMPL_ID);

                SW_NO_MORE_SEGS = '0'B;

                DO WHILE (¬SW_END_OF_DB & ¬SW_NO_MORE_SEGS);
                   CALL P500_GET_PAY_SEG;
                END; /* DO WHILE */

                 /* GET NEXT ROOT */

                CALL PLITDLI (FOUR,
                              DLI_FUNCGN,
                              PCB_MASK,
                              IO_EMPLOYEE_RECORD,
                              EMP_UNQUALIFIED_SSA);

                IF STATUS_CODE = '  ' THEN;
                ELSE
                   IF STATUS_CODE = 'GB' THEN
                      DO;
                          SW_END_OF_DB = '1'B;
                          PUT SKIP LIST ('** END OF DATABASE');
                      END;
                   ELSE
                      DO;
                          CALL P400_DISPLAY_ERROR;
                          RETURN;
                      END;
```

```
              END; /* DO WHILE */

           PUT SKIP LIST ('FINISHED PROCESSING IN P200_MAINLINE');

END P200_MAINLINE;

P300_TERMINATION: PROC;

    CLOSE FILE(EMPIFILE);

    PUT SKIP LIST ('PLIIMS7 - SUCCESSFULLY ENDED');

END P300_TERMINATION;

P400_DISPLAY_ERROR: PROC;

    PUT SKIP LIST ('ERROR ENCOUNTERED - DETAIL FOLLOWS');
    PUT SKIP LIST ('SEG_IO_AREA      :' || SEG_IO_AREA);
    PUT SKIP LIST ('DBD_NAME1:' ||  DBD_NAME);
    PUT SKIP LIST ('SEG_LEVEL1:' || SEG_LEVEL);
    PUT SKIP LIST ('STATUS_CODE:' || STATUS_CODE);
    PUT SKIP LIST ('PROC_OPT1 :' || PROC_OPT);
    PUT SKIP LIST ('SEG_NAME1 :' || SEG_NAME);
    PUT SKIP LIST ('KEY_FDBK1 :' || KEY_FDBK);
    PUT SKIP LIST ('NUM_SENSEG1:' || NUM_SENSEG);
    PUT SKIP LIST ('KEY_FDBK_AREA1:' || KEY_FDBK_AREA);

END P400_DISPLAY_ERROR;

P500_GET_PAY_SEG: PROC;

    PUT SKIP LIST ('PROCESSING IN P500_GET_PAY_SEG');
    EMP_ID_VAL  = EMPL_ID;
    EFFDATE_VAL = '20170101';

    CALL PLITDLI (FIVE,
                  DLI_FUNCGNP,
                  PCB_MASK,
                  IO_EMPPAY_RECORD,
                  EMP_QUALIFIED_SSA,
                  EMPPAY_QUALIFIED_SSA);

    SELECT (STATUS_CODE);

       WHEN ('  ')
          DO;
             PUT SKIP LIST ('SUCCESSFUL EMPPAY RETRIEVAL');
             PUT SKIP LIST ('PAY_EFF_DATE ' || PAY_EFF_DATE);
             PUT SKIP LIST ('PAY_REG_PAY  ' || PAY_REG_PAY);
             PUT SKIP LIST ('PAY_BON_PAY  ' || PAY_BON_PAY);
             PUT SKIP LIST ('SEMIMTH_PAY  ' || SEMIMTH_PAY);
          END;

       WHEN ('GE')
```

93

```
        DO;
            SW_NO_MORE_SEGS = '1'B;
            PUT SKIP LIST ('** NO MORE PAY SEGMENTS');
        END;

    WHEN ('GB')
        DO;
            SW_END_OF_DB = '1'B;
            PUT SKIP LIST ('** END OF DATABASE');
        END;
    OTHERWISE
        CALL P400_DISPLAY_ERROR;

    END; /* SELECT */

END P500_GET_PAY_SEG;

END PLIIMS7;
```

Inserting Child Segments Down the Hierarchy (3 levels)

Ok, I think we have a pretty good handle on the adding and retrieving of child segments. But just to be sure, let's work with the EMPPAYHS segment, adding and retrieving records. That's slightly different that what we've done already, but not much.

For COBIMS8, let's add a pay history segment EMPPAYHS for all employees using pay date January 15, 2017, and using the twice-monthly pay information from the EMPPAY segment. So we need to position ourselves at the EMPPAY child segment under each EMPLOYEE root segment, and then ISRT an EMPPAYHS segment.

I think we've covered all the techniques required to write this program. Why don't you give it a try first, and then we'll get back together and compare our code? Take a good break and then code up your version.

. . . .

Ok, I'm back with a good cup of coffee. Here's my version of the code. I added the segment I/O and SSAs for the EMPPAYHS segment. The INSERT call for the EMPPAY-HS segment is as follows:

```
        CALL 'CBLTDLI' USING SIX,
                DLI-FUNCISRT,
                PCB-MASK,
                IO-EMPPAYHS-RECORD,
                EMP-QUALIFIED-SSA,
```

```
          EMPPAY-QUALIFIED-SSA,
          EMPPAYHS-UNQUALIFIED-SSA.
```

Should be no surprises there. Just a bit more navigation and slightly different database calls. Note that we must use qualified SSAs for EMPLOYEE and EMPPAY. Here's the full program.

```
      IDENTIFICATION DIVISION.
      PROGRAM-ID. COBIMS8.
      *********************************************************
      *    INSERT EMPLOYEE PAY HISTORY RECS INTO THE        *
      *    EMPLOYEE IMS DATABASE. THIS EXAMPLE WALKS         *
      *    THROUGH THE ROOT AND EMPPAY SEGS AND THEN         *
      *    INSERTS THE PAY HISTORY SEGMENT UNDER THE         *
      *    EMPPAY SEGMENT.                                   *
      *********************************************************
      ENVIRONMENT DIVISION.
      DATA DIVISION.

      *********************************************************
      * W O R K I N G   S T O R A G E   S E C T I O N   *
      *********************************************************
      WORKING-STORAGE SECTION.

      01 WS-FLAGS.
         05  SW-END-OF-FILE-SWITCH   PIC X(1) VALUE 'N'.
             88  SW-END-OF-FILE                VALUE 'Y'.
             88  SW-NOT-END-OF-FILE            VALUE 'N'.
         05  SW-END-OF-DB-SWITCH     PIC X(1) VALUE 'N'.
             88  SW-END-OF-DB                  VALUE 'Y'.
             88  SW-NOT-END-OF-DB              VALUE 'N'.

      01 IO-EMPLOYEE-RECORD.
         05  EMPL-ID       PIC X(04).
         05  FILLER        PIC X(01).
         05  EMPL-LNAME    PIC X(30).
         05  FILLER        PIC X(01).
         05  EMPL-FNAME    PIC X(20).
         05  FILLER        PIC X(01).
         05  EMPL-YRS-SRV  PIC X(02).
         05  FILLER        PIC X(01).
         05  EMPL-PRM-DTE  PIC X(10).
         05  FILLER        PIC X(10).
```

```
01 IO-EMPPAY-RECORD.
   05  PAY-EFF-DATE  PIC X(8).
   05  PAY-REG-PAY   PIC S9(6)V9(2) USAGE COMP-3.
   05  PAY-BON-PAY   PIC S9(6)V9(2) USAGE COMP-3.
   05  SEMIMTH-PAY   PIC S9(6)V9(2) USAGE COMP-3.
   05  FILLER        PIC X(57).

01 IO-EMPPAYHS-RECORD.
   05  PAY-DATE      PIC X(8).
   05  PAY-ANN-PAY   PIC S9(6)V9(2) USAGE COMP-3.
   05  PAY-AMT       PIC S9(6)V9(2) USAGE COMP-3.
   05  FILLER        PIC X(62).

01 SEG-IO-AREA    PIC X(80).
01 DLI-FUNCTIONS.
   05 DLI-FUNCISRT PIC X(4) VALUE 'ISRT'.
   05 DLI-FUNCGU   PIC X(4) VALUE 'GU  '.
   05 DLI-FUNCGN   PIC X(4) VALUE 'GN  '.
   05 DLI-FUNCGHU  PIC X(4) VALUE 'GHU '.
   05 DLI-FUNCGNP  PIC X(4) VALUE 'GNP '.
   05 DLI-FUNCREPL PIC X(4) VALUE 'REPL'.
   05 DLI-FUNCDLET PIC X(4) VALUE 'DLET'.
   05 DLI-FUNCXRST PIC X(4) VALUE 'XRST'.
   05 DLI-FUNCCKPT PIC X(4) VALUE 'CKPT'.

 01 EMP-UNQUALIFIED-SSA.
    05  SEGNAME     PIC X(08) VALUE 'EMPLOYEE'.
    05  FILLER      PIC X(01) VALUE ' '.

 01 EMP-QUALIFIED-SSA.
    05  SEGNAME     PIC X(08) VALUE 'EMPLOYEE'.
    05  FILLER      PIC X(01) VALUE '('.
    05  FIELD       PIC X(08) VALUE 'EMPID'.
    05  OPER        PIC X(02) VALUE ' ='.
    05  EMP-ID-VAL  PIC X(04) VALUE '    '.
    05  FILLER      PIC X(01) VALUE ')'.

 01 EMPPAY-UNQUALIFIED-SSA.
    05  SEGNAME     PIC X(08) VALUE 'EMPPAY  '.
    05  FILLER      PIC X(01) VALUE ' '.

 01 EMPPAY-QUALIFIED-SSA.
    05  SEGNAME     PIC X(08) VALUE 'EMPPAY  '.
    05  FILLER      PIC X(01) VALUE '('.
    05  FIELD       PIC X(08) VALUE 'EFFDATE '.
    05  OPER        PIC X(02) VALUE ' ='.
    05  EFFDATE-VAL PIC X(08) VALUE '        '.
    05  FILLER      PIC X(01) VALUE ')'.
```

```cobol
    01  EMPPAYHS-UNQUALIFIED-SSA.
        05  SEGNAME      PIC X(08) VALUE 'EMPPAYHS'.
        05  FILLER       PIC X(01) VALUE ' '.


    01  IMS-RET-CODES.
        05  THREE        PIC S9(9) COMP VALUE +3.
        05  FOUR         PIC S9(9) COMP VALUE +4.
        05  FIVE         PIC S9(9) COMP VALUE +5.
        05  SIX          PIC S9(9) COMP VALUE +6.

    77  WS-PAY-DATE      PIC X(08) VALUE '20170115'.

    LINKAGE SECTION.
     01  PCB-MASK.
        03  DBD-NAME     PIC X(8).
        03  SEG-LEVEL    PIC XX.
        03  STATUS-CODE  PIC XX.
        03  PROC-OPT     PIC X(4).
        03  FILLER       PIC X(4).
        03  SEG-NAME     PIC X(8).
        03  KEY-FDBK     PIC S9(5) COMP.
        03  NUM-SENSEG   PIC S9(5) COMP.
        03  KEY-FDBK-AREA.
           05  EMPLOYEE-ID  PIC X(04).
           05  EMPPAYHS     PIC X(08).

    PROCEDURE DIVISION.

        INITIALIZE PCB-MASK
        ENTRY 'DLITCBL' USING PCB-MASK

        PERFORM P100-INITIALIZATION.
        PERFORM P200-MAINLINE.
        PERFORM P300-TERMINATION.
        GOBACK.

    P100-INITIALIZATION.

        DISPLAY '** PROGRAM COBIMS8 START **'
        DISPLAY 'PROCESSING IN P100-INITIALIZATION'.

*       DO INITIAL DB READ FOR FIRST EMPLOYEE ROOT SEGMENT

        CALL 'CBLTDLI' USING ,
             DLI-FUNCGN,
             PCB-MASK,
```

```
                IO-EMPLOYEE-RECORD,
                EMP-UNQUALIFIED-SSA

        IF STATUS-CODE = '  ' THEN
            NEXT SENTENCE
        ELSE
            IF STATUS-CODE = 'GB' THEN
                SET SW-END-OF-DB TO TRUE
                DISPLAY 'END OF DATABASE :'
            ELSE
                PERFORM P9000-DISPLAY-ERROR
                GOBACK
            END-IF

        END-IF.

    P200-MAINLINE.
        DISPLAY 'PROCESSING IN P200-MAINLINE'

*       CHECK STATUS CODE AND FIRST RECORD

        IF SW-END-OF-DB THEN
            DISPLAY 'NO RECORDS TO PROCESS!!'
        ELSE
            PERFORM UNTIL SW-END-OF-DB
                DISPLAY 'SUCCESSFUL READ :' IO-EMPLOYEE-RECORD
                MOVE EMPL-ID TO EMP-ID-VAL
                PERFORM P2000-GET-EMPPAY
                IF STATUS-CODE NOT EQUAL SPACES THEN
                    PERFORM P9000-DISPLAY-ERROR
                    GOBACK
                ELSE
                    DISPLAY 'SUCCESSFUL PAY READ :' IO-EMPPAY-RECORD
                    MOVE PAY-EFF-DATE TO EFFDATE-VAL
                    MOVE WS-PAY-DATE TO PAY-DATE
                    MOVE PAY-REG-PAY TO PAY-ANN-PAY
                    MOVE SEMIMTH-PAY TO PAY-AMT
                    PERFORM P3000-INSERT-EMPPAYHS
                    IF STATUS-CODE NOT EQUAL SPACES THEN
                        PERFORM P9000-DISPLAY-ERROR
                        GOBACK
                    ELSE
                        DISPLAY 'SUCCESSFUL INSERT EMPPAYHS : '
                            EMP-ID-VAL
                        DISPLAY 'SUCCESSFUL INSERT VALUES   : '
                            IO-EMPPAYHS-RECORD
                    END-IF
```

```
            PERFORM P1000-GET-NEXT-ROOT
            IF STATUS-CODE = 'GB' THEN
                SET SW-END-OF-DB TO TRUE
                DISPLAY 'END OF DATABASE'
            END-IF

        END-IF

    END-PERFORM.

    DISPLAY 'FINISHED PROCESSING IN P200-MAINLINE'.

P300-TERMINATION.

    DISPLAY 'PROCESSING IN P300-TERMINATION'
    DISPLAY '** COBIMS8 - SUCCESSFULLY ENDED **'.

P1000-GET-NEXT-ROOT.

    DISPLAY 'PROCESSING IN P1000-GET-NEXT-ROOT'.

    CALL 'CBLTDLI' USING FOUR,
            DLI-FUNCGN,
            PCB-MASK,
            IO-EMPLOYEE-RECORD,
            EMP-UNQUALIFIED-SSA.

P2000-GET-EMPPAY.

    DISPLAY 'PROCESSING IN P2000-GET-EMPPAY'.

    CALL 'CBLTDLI' USING FIVE,
            DLI-FUNCGNP,
            PCB-MASK,
            IO-EMPPAY-RECORD,
            EMP-QUALIFIED-SSA,
            EMPPAY-UNQUALIFIED-SSA.

P3000-INSERT-EMPPAYHS.

    DISPLAY 'PROCESSING IN P3000-INSERT-EMPPAYHS'.

    CALL 'CBLTDLI' USING SIX,
            DLI-FUNCISRT,
            PCB-MASK,
            IO-EMPPAYHS-RECORD,
            EMP-QUALIFIED-SSA,
            EMPPAY-QUALIFIED-SSA,
```

```
              EMPPAYHS-UNQUALIFIED-SSA.

      P9000-DISPLAY-ERROR.

          DISPLAY 'ERROR ENCOUNTERED - DETAIL FOLLOWS'
          DISPLAY 'DBD-NAME1:'      DBD-NAME
          DISPLAY 'SEG-LEVEL1:'     SEG-LEVEL
          DISPLAY 'STATUS-CODE:'    STATUS-CODE
          DISPLAY 'PROC-OPT1 :'     PROC-OPT
          DISPLAY 'SEG-NAME1 :'     SEG-NAME
          DISPLAY 'KEY-FDBK1 :'     KEY-FDBK
          DISPLAY 'NUM-SENSEG1:'    NUM-SENSEG
          DISPLAY 'KEY-FDBK-AREA1:' KEY-FDBK-AREA.

      *    END OF SOURCE CODE
```

Now let's compile, link and run the program. Here is the output.

```
** PROGRAM COBIMS8 START **
PROCESSING IN P100-INITIALIZATION
PROCESSING IN P200-MAINLINE
SUCCESSFUL READ :1111 VEREEN                 CHARLES              12 201
PROCESSING IN P2000-GET-EMPPAY
SUCCESSFUL PAY READ :20170101 g          &
PROCESSING IN P3000-INSERT-EMPPAYHS
SUCCESSFUL INSERT EMPPAYHS : 1111
SUCCESSFUL INSERT VALUES   : 20170115 g      &
PROCESSING IN P1000-GET-NEXT-ROOT
SUCCESSFUL READ :1122 JENKINS                DEBORAH              05 201
PROCESSING IN P2000-GET-EMPPAY
SUCCESSFUL PAY READ :20170101 b           %
PROCESSING IN P3000-INSERT-EMPPAYHS
SUCCESSFUL INSERT EMPPAYHS : 1122
SUCCESSFUL INSERT VALUES   : 20170115 b       %
PROCESSING IN P1000-GET-NEXT-ROOT
SUCCESSFUL READ :3217 JOHNSON                EDWARD               04 201
PROCESSING IN P2000-GET-EMPPAY
SUCCESSFUL PAY READ :20170101        &    c
PROCESSING IN P3000-INSERT-EMPPAYHS
SUCCESSFUL INSERT EMPPAYHS : 3217
SUCCESSFUL INSERT VALUES   : 20170115        c
PROCESSING IN P1000-GET-NEXT-ROOT
SUCCESSFUL READ :4175 TURNBULL               FRED                 01 201
PROCESSING IN P2000-GET-EMPPAY
SUCCESSFUL PAY READ :20170101        &    %
PROCESSING IN P3000-INSERT-EMPPAYHS
SUCCESSFUL INSERT EMPPAYHS : 4175
SUCCESSFUL INSERT VALUES   : 20170115        %
PROCESSING IN P1000-GET-NEXT-ROOT
SUCCESSFUL READ :4720 SCHULTZ                TIM                  09 201
PROCESSING IN P2000-GET-EMPPAY
SUCCESSFUL PAY READ :20170101       &
```

```
PROCESSING IN P3000-INSERT-EMPPAYHS
SUCCESSFUL INSERT EMPPAYHS : 4720
SUCCESSFUL INSERT VALUES   : 20170115
PROCESSING IN P1000-GET-NEXT-ROOT
SUCCESSFUL READ :4836 SMITH                      SANDRA           03 201
PROCESSING IN P2000-GET-EMPPAY
SUCCESSFUL PAY READ :20170101
PROCESSING IN P3000-INSERT-EMPPAYHS
SUCCESSFUL INSERT EMPPAYHS : 4836
SUCCESSFUL INSERT VALUES   : 20170115
PROCESSING IN P1000-GET-NEXT-ROOT
SUCCESSFUL READ :6288 WILLARD                    JOE              06 201
PROCESSING IN P2000-GET-EMPPAY
SUCCESSFUL PAY READ :20170101            j %
PROCESSING IN P3000-INSERT-EMPPAYHS
SUCCESSFUL INSERT EMPPAYHS : 6288
SUCCESSFUL INSERT VALUES   : 20170115      j %
PROCESSING IN P1000-GET-NEXT-ROOT
SUCCESSFUL READ :7459 STEWART                     BETTY           07 201
PROCESSING IN P2000-GET-EMPPAY
SUCCESSFUL PAY READ :20170101 e       &        %
PROCESSING IN P3000-INSERT-EMPPAYHS
SUCCESSFUL INSERT EMPPAYHS : 7459
SUCCESSFUL INSERT VALUES   : 20170115 e         %
PROCESSING IN P1000-GET-NEXT-ROOT
END OF DATABASE
FINISHED PROCESSING IN P200-MAINLINE
PROCESSING IN P300-TERMINATION
** COBIMS8 - SUCCESSFULLY ENDED **
```

So that's how to insert a child segment under a higher level child. To make this more interesting, change the value of the pay date to January 31, 2017. Then compile and link and run again. Do this twice more using pay dates February 15, 2017 and February 28, 2017. Now we have four paychecks for each employee. We'll read all this data back in the next training program.

Here is the PLI code that corresponds to COBIMS8.

```
PLIIMS8: PROCEDURE (DB_PTR_PCB) OPTIONS(MAIN);

  /*****************************************************************
  * PROGRAM NAME :   PLIIMS8 - INSERT EMPLOYEE HISTORY PAY RECORDS  *
  *                  UNDER THEIR EMPPAY PARENTS. QUALIFIED SSA'S    *
  *                  MUST BE PROVIDED FOR BOTH THE EMPLOYEE AND     *
  *                  EMPPAY SEGMENTS.                               *
  *****************************************************************/
  /*****************************************************************
  /*            W O R K I N G   S T O R A G E                    *
  *****************************************************************/
```

```
DCL SW_END_OF_DB              STATIC BIT(01) INIT('0'B);
DCL ONCODE                    BUILTIN;
DCL DB_PTR_PCB                POINTER;
DCL PLITDLI                   EXTERNAL ENTRY;

DCL 01 DLI_FUNCTIONS,
       05 DLI_FUNCISRT        CHAR(04) INIT ('ISRT'),
       05 DLI_FUNCGU          CHAR(04) INIT ('GU  '),
       05 DLI_FUNCGN          CHAR(04) INIT ('GN  '),
       05 DLI_FUNCGHU         CHAR(04) INIT ('GHU '),
       05 DLI_FUNCGNP         CHAR(04) INIT ('GNP '),
       05 DLI_FUNCREPL        CHAR(04) INIT ('REPL'),
       05 DLI_FUNCDLET        CHAR(04) INIT ('DLET'),
       05 DLI_FUNCXRST        CHAR(04) INIT ('XRST'),
       05 DLI_FUNCCHKP        CHAR(04) INIT ('CHKP'),
       05 DLI_FUNCROLL        CHAR(04) INIT ('ROLL');

DCL 01 IO_EMPLOYEE_RECORD,
       05  EMPL_ID            CHAR(04),
       05  FILLER1            CHAR(01),
       05  EMPL_LNAME         CHAR(30),
       05  FILLER2            CHAR(01),
       05  EMPL_FNAME         CHAR(20),
       05  FILLER3            CHAR(01),
       05  EMPL_YRS_SRV       CHAR(02),
       05  FILLER4            CHAR(01),
       05  EMPL_PRM_DTE       CHAR(10),
       05  FILLER5            CHAR(10);

DCL 01 IO_EMPPAY_RECORD,
       05  PAY_EFF_DATE       CHAR(8),
       05  PAY_REG_PAY        FIXED DEC (8,2),
       05  PAY_BON_PAY        FIXED DEC (8,2),
       05  SEMIMTH_PAY        FIXED DEC (8,2),
       05  FILLER55           CHAR(57);

DCL 01 IO_EMPPAYHS_RECORD,
       05  PAY_DATE           CHAR(8),
       05  PAY_ANN_PAY        FIXED DEC (8,2),
       05  PAY_AMT            FIXED DEC (8,2),
       05  FILLER65           CHAR(62);

DCL 01 PCB_MASK              BASED(DB_PTR_PCB),
       05  DBD_NAME           CHAR(08),
       05  SEG_LEVEL          CHAR(02),
       05  STATUS_CODE        CHAR(02),
       05  PROC_OPT           CHAR(04),
       05  FILLER6            FIXED BIN (31),
       05  SEG_NAME           CHAR(08),
       05  KEY_FDBK           FIXED BIN (31),
       05  NUM_SENSEG         FIXED BIN (31),
       05  KEY_FDBK_AREA,
           10 EMPLOYEE_ID     CHAR(04),
           10 EMP_PAY_DATE    CHAR(08);
```

```
       DCL 01 EMP_UNQUALIFIED_SSA,
              05  SEGNAME            CHAR(08) INIT ('EMPLOYEE'),
              05  FILLER7            CHAR(01) INIT (' ');

       DCL 01 EMP_QUALIFIED_SSA,
              05  SEGNAME            CHAR(08) INIT('EMPLOYEE'),
              05  FILLER8            CHAR(01) INIT('('),
              05  FIELD              CHAR(08) INIT('EMPID'),
              05  OPER               CHAR(02) INIT(' ='),
              05  EMP_ID_VAL         CHAR(04) INIT('    '),
              05  FILLER9            CHAR(01) INIT(')');

       DCL 01 EMPPAY_UNQUALIFIED_SSA,
              05  SEGNAME            CHAR(08) INIT('EMPPAY  '),
              05  FILLER10           CHAR(01) INIT(' ');

       DCL 01 EMPPAY_QUALIFIED_SSA,
              05  SEGNAME            CHAR(08) INIT('EMPPAY  '),
              05  FILLER11           CHAR(01) INIT('('),
              05  FIELD              CHAR(08) INIT('EFFDATE '),
              05  OPER               CHAR(02) INIT(' ='),
              05  EFFDATE_VAL        CHAR(08) INIT('        '),
              05  FILLER12           CHAR(01) INIT(')');

       DCL 01 EMPPAYHS_UNQUALIFIED_SSA,
              05  SEGNAME            CHAR(08) INIT('EMPPAYHS'),
              05  FILLER13           CHAR(01) INIT(' ');

       DCL WS_PAY_DATE               CHAR(08) INIT ('20170228');

       DCL SEG_IO_AREA               CHAR(80) INIT (' ');

       DCL THREE                     FIXED BIN (31) INIT(3);
       DCL FOUR                      FIXED BIN (31) INIT(4);
       DCL FIVE                      FIXED BIN (31) INIT(5);
       DCL SIX                       FIXED BIN (31) INIT(6);

/*********************************************************************
/*               P R O G R A M   M A I N L I N E               *
*********************************************************************/

CALL P100_INITIALIZATION;
CALL P200_MAINLINE;
CALL P300_TERMINATION;

P100_INITIALIZATION: PROC;

   PUT SKIP LIST ('PLIIMS8: INSERT RECORDS');
   PUT SKIP LIST ('PROCESSING IN P100_INITIALIZATION');

   IO_EMPLOYEE_RECORD  = '';
   IO_EMPPAY_RECORD   = '';
   IO_EMPPAYHS_RECORD = '';
```

103

```
        PCB_MASK = '';

        /* DO INITIAL DB READ FOR FIRST EMPLOYEE RECORD */

        CALL PLITDLI (FOUR,
                      DLI_FUNCGN,
                      PCB_MASK,
                      IO_EMPLOYEE_RECORD,
                      EMP_UNQUALIFIED_SSA);

        IF STATUS_CODE = '  ' THEN;
        ELSE
           IF STATUS_CODE = 'GB' THEN
              DO;
                 SW_END_OF_DB = '1'B;
                 PUT SKIP LIST ('** END OF DATABASE');
              END;
           ELSE
              DO;
                 CALL P9000_DISPLAY_ERROR;
                 RETURN;
              END;

END P100_INITIALIZATION;

P200_MAINLINE: PROC;

    /*  MAIN LOOP - WALK THROUGH THE DATABASE GETTING EMPLOYEE
                    PAY HISTORY SEGMENTS.                      */

    IF SW_END_OF_DB THEN
       PUT SKIP LIST ('NO RECORDS TO PROCESS!!');
    ELSE
       DO WHILE (¬SW_END_OF_DB);

          PUT SKIP LIST ('SUCCESSFUL READ OF EMPLOYEE ' || EMPL_ID);

          /* ASSIGN KEY FOR EMPLOYEE LEVEL SSA AND GET EMPPAY SEG */

          EMP_ID_VAL = EMPL_ID;
          CALL P2000_GET_EMPPAY;
          IF STATUS_CODE ¬= '  ' THEN
             DO;
                PUT SKIP LIST ('ERROR READING EMPPAY');
                CALL P9000_DISPLAY_ERROR;
                RETURN;
             END;
          ELSE
             DO;
                PUT SKIP LIST ('SUCCESSFUL PAY READ : '
                   || IO_EMPPAY_RECORD);

                /* ASSIGN KEY FOR EMPPAY LEVEL SSA FOR SEG */
```

```
                    EFFDATE_VAL = PAY_EFF_DATE;

                    PAY_DATE    = WS_PAY_DATE;
                    PAY_ANN_PAY = PAY_REG_PAY;
                    PAY_AMT     = SEMIMTH_PAY;

                    CALL P3000_INSERT_EMPPAYHS;
                    IF STATUS_CODE ¬= ' ' THEN
                        DO;
                            CALL P9000_DISPLAY_ERROR;
                            RETURN;
                        END;
                    ELSE
                        DO;
                            PUT SKIP LIST ('SUCCESSFUL INSERT EMPPAYHS : '
                                || EMP_ID_VAL);
                            PUT SKIP LIST ('SUCCESSFUL INSERT VALUES   : ' );
                            PUT SKIP DATA (IO_EMPPAYHS_RECORD);
                        END;

                    CALL P1000_GET_NEXT_ROOT;
                    IF STATUS_CODE = 'GB' THEN
                        DO;
                            SW_END_OF_DB = '1'B;
                            PUT SKIP LIST ('** END OF DATABASE');
                        END;
                    ELSE
                        IF STATUS_CODE = ' ' THEN;
                        ELSE
                            DO;
                                CALL P9000_DISPLAY_ERROR;
                                RETURN;
                            END;

            END; /* SUCCESSFULLY RETRIEVED PAY SEG */

        END; /* DO WHILE */

    PUT SKIP LIST ('FINISHED PROCESSING IN P200_MAINLINE');

END P200_MAINLINE;

P300_TERMINATION: PROC;

    PUT SKIP LIST ('PLIIMS8 - SUCCESSFULLY ENDED');

END P300_TERMINATION;

P1000_GET_NEXT_ROOT: PROC;

    PUT SKIP LIST ('PROCESSING IN P1000_GET_NEXT_ROOT');

    CALL PLITDLI (FOUR,
                  DLI_FUNCGN,
```

```
                              PCB_MASK,
                              IO_EMPLOYEE_RECORD,
                              EMP_UNQUALIFIED_SSA);

END P1000_GET_NEXT_ROOT;

P2000_GET_EMPPAY: PROC;

   PUT SKIP LIST ('PROCESSING IN P2000_GET_EMPPAY');

   CALL PLITDLI (FIVE,
                 DLI_FUNCGNP,
                 PCB_MASK,
                 IO_EMPPAY_RECORD,
                 EMP_QUALIFIED_SSA,
                 EMPPAY_UNQUALIFIED_SSA);

END P2000_GET_EMPPAY;

P3000_INSERT_EMPPAYHS: PROC;

   PUT SKIP LIST ('PROCESSING IN P3000-INSERT-EMPPAYHS');

   CALL PLITDLI (SIX,
                 DLI_FUNCISRT,
                 PCB_MASK,
                 IO_EMPPAYHS_RECORD,
                 EMP_QUALIFIED_SSA,
                 EMPPAY_QUALIFIED_SSA,
                 EMPPAYHS_UNQUALIFIED_SSA);

END P3000_INSERT_EMPPAYHS;

P9000_DISPLAY_ERROR: PROC;

   PUT SKIP LIST ('ERROR ENCOUNTERED - DETAIL FOLLOWS');
   PUT SKIP DATA (IO_EMPPAY_RECORD);
   PUT SKIP LIST ('DBD_NAME1:' || DBD_NAME);
   PUT SKIP LIST ('SEG_LEVEL1:' || SEG_LEVEL);
   PUT SKIP LIST ('STATUS_CODE:' || STATUS_CODE);
   PUT SKIP LIST ('PROC_OPT1 :' || PROC_OPT);
   PUT SKIP LIST ('SEG_NAME1 :' || SEG_NAME);
   PUT SKIP LIST ('KEY_FDBK1 :' || KEY_FDBK);
   PUT SKIP LIST ('NUM_SENSEG1:' || NUM_SENSEG);
   PUT SKIP LIST ('KEY_FDBK_AREA1:' || KEY_FDBK_AREA);

END P9000_DISPLAY_ERROR;

END PLIIMS8;
```

Read Child Segments Down the Hierarchy (3 levels)

For COBIMS9 you'll need to retrieve and display all the pay history segments for each employee. This should be fairly straightforward by now. Yes you'll need one more loop, and more navigation. But we need the practice to really drill the techniques in. Give this one a try, then take a long break and we'll compare code.

. . . .

Ok, I hope you are enjoying coding IMS in COBOL! I'll bet you got your version of the program to work without any serious problems. Let me give you my code and see what you think. Note that I have switches both for end of database and for end of EMPPAYHS segments. The latter is needed for looping through the multiple EMPPAYHS segments.

```
       IDENTIFICATION DIVISION.
       PROGRAM-ID. COBIMS9.
      *********************************************************
      *    READ AND DISPLAY EMP HISTORY RECS FROM THE        *
      *    EMPLOYEE IMS DATABASE. THIS EXAMPLE WALKS          *
      *    THROUGH THE ROOT AND EMPPAY SEGS AND THEN          *
      *    READS THE PAY HISTORY SEGMENTS UNDER THE           *
      *    EMPPAY SEGMENT.                                    *
      *********************************************************
       ENVIRONMENT DIVISION.
       DATA DIVISION.
      *********************************************************
      * W O R K I N G   S T O R A G E   S E C T I O N   *
      *********************************************************
       WORKING-STORAGE SECTION.
       01 WS-FLAGS.
          05  SW-END-OF-FILE-SWITCH   PIC X(1) VALUE 'N'.
              88  SW-END-OF-FILE                VALUE 'Y'.
              88  SW-NOT-END-OF-FILE            VALUE 'N'.
          05  SW-END-OF-DB-SWITCH     PIC X(1) VALUE 'N'.
              88  SW-END-OF-DB                  VALUE 'Y'.
              88  SW-NOT-END-OF-DB              VALUE 'N'.
          05  SW-END-OF-EMPPAYHS-SW   PIC X(1) VALUE 'N'.
              88  SW-END-OF-EMPPAYHS            VALUE 'Y'.
              88  SW-NOT-END-OF-EMPPAYHS        VALUE 'N'.

       01 IO-EMPLOYEE-RECORD.
          05  EMPL-ID      PIC X(04).
          05  FILLER       PIC X(01).
          05  EMPL-LNAME   PIC X(30).
```

107

```
      05   FILLER         PIC X(01).
      05   EMPL-FNAME     PIC X(20).
      05   FILLER         PIC X(01).
      05   EMPL-YRS-SRV   PIC X(02).
      05   FILLER         PIC X(01).
      05   EMPL-PRM-DTE   PIC X(10).
      05   FILLER         PIC X(10).

01 IO-EMPPAY-RECORD.
      05   PAY-EFF-DATE   PIC X(8).
      05   PAY-REG-PAY    PIC S9(6)V9(2) USAGE COMP-3.
      05   PAY-BON-PAY    PIC S9(6)V9(2) USAGE COMP-3.
      05   SEMIMTH-PAY    PIC S9(6)V9(2) USAGE COMP-3.
      05   FILLER         PIC X(57).

01 IO-EMPPAYHS-RECORD.
      05   PAY-DATE       PIC X(8).
      05   PAY-ANN-PAY    PIC S9(6)V9(2) USAGE COMP-3.
      05   PAY-AMT        PIC S9(6)V9(2) USAGE COMP-3.
      05   FILLER         PIC X(62).

01 SEG-IO-AREA     PIC X(80).

01 DLI-FUNCTIONS.
      05 DLI-FUNCISRT  PIC X(4) VALUE 'ISRT'.
      05 DLI-FUNCGU    PIC X(4) VALUE 'GU  '.
      05 DLI-FUNCGN    PIC X(4) VALUE 'GN  '.
      05 DLI-FUNCGHU   PIC X(4) VALUE 'GHU '.
      05 DLI-FUNCGNP   PIC X(4) VALUE 'GNP '.
      05 DLI-FUNCREPL  PIC X(4) VALUE 'REPL'.
      05 DLI-FUNCDLET  PIC X(4) VALUE 'DLET'.
      05 DLI-FUNCXRST  PIC X(4) VALUE 'XRST'.
      05 DLI-FUNCCKPT  PIC X(4) VALUE 'CKPT'.

01 DISPLAY-EMPPAYHS-PIC.
      05  DIS-REG-PAY   PIC ZZ999.99-.
      05  DIS-SMT-PAY   PIC ZZ999.99-.

 01 EMP-UNQUALIFIED-SSA.
      05  SEGNAME      PIC X(08) VALUE 'EMPLOYEE'.
      05  FILLER       PIC X(01) VALUE ' '.

 01 EMP-QUALIFIED-SSA.
      05  SEGNAME      PIC X(08) VALUE 'EMPLOYEE'.
      05  FILLER       PIC X(01) VALUE '('.
      05  FIELD        PIC X(08) VALUE 'EMPID'.
      05  OPER         PIC X(02) VALUE ' ='.
      05  EMP-ID-VAL   PIC X(04) VALUE '    '.
```

```cobol
        05  FILLER      PIC X(01) VALUE ')'.

    01 EMPPAY-UNQUALIFIED-SSA.
        05  SEGNAME     PIC X(08) VALUE 'EMPPAY  '.
        05  FILLER      PIC X(01) VALUE ' '.

    01 EMPPAY-QUALIFIED-SSA.
        05  SEGNAME     PIC X(08) VALUE 'EMPPAY  '.
        05  FILLER      PIC X(01) VALUE '('.
        05  FIELD       PIC X(08) VALUE 'EFFDATE '.
        05  OPER        PIC X(02) VALUE ' ='.
        05  EFFDATE-VAL PIC X(08) VALUE '        '.
        05  FILLER      PIC X(01) VALUE ')'.

    01 EMPPAYHS-UNQUALIFIED-SSA.
        05  SEGNAME     PIC X(08) VALUE 'EMPPAYHS'.
        05  FILLER      PIC X(01) VALUE ' '.

    01 IMS-RET-CODES.
        05 THREE        PIC S9(9) COMP VALUE +3.
        05 FOUR         PIC S9(9) COMP VALUE +4.
        05 FIVE         PIC S9(9) COMP VALUE +5.
        05 SIX          PIC S9(9) COMP VALUE +6.

    77 WS-PAY-DATE      PIC X(08) VALUE '20170228'.

LINKAGE SECTION.
 01 PCB-MASK.
        03 DBD-NAME      PIC X(8).
        03 SEG-LEVEL     PIC XX.
        03 STATUS-CODE   PIC XX.
        03 PROC-OPT      PIC X(4).
        03 FILLER        PIC X(4).
        03 SEG-NAME      PIC X(8).
        03 KEY-FDBK      PIC S9(5) COMP.
        03 NUM-SENSEG    PIC S9(5) COMP.
        03 KEY-FDBK-AREA.
           05 EMPLOYEE-ID  PIC X(04).
           05 EMPPAYHS     PIC X(08).

PROCEDURE DIVISION.

    INITIALIZE PCB-MASK
    ENTRY 'DLITCBL' USING PCB-MASK

    PERFORM P100-INITIALIZATION.
    PERFORM P200-MAINLINE.
    PERFORM P300-TERMINATION.
```

```
        GOBACK.

    P100-INITIALIZATION.

        DISPLAY '** PROGRAM COBIMS9 START **'
        DISPLAY 'PROCESSING IN P100-INITIALIZATION'.

*       DO INITIAL DB READ FOR FIRST EMPLOYEE ROOT SEGMENT

        CALL 'CBLTDLI' USING FOUR,
             DLI-FUNCGN,
             PCB-MASK,
             IO-EMPLOYEE-RECORD,
             EMP-UNQUALIFIED-SSA

        IF STATUS-CODE = '  ' THEN
            DISPLAY '********************************'
        ELSE
            IF STATUS-CODE = 'GB' THEN
                SET SW-END-OF-DB TO TRUE
                DISPLAY 'END OF DATABASE :'
            ELSE
                PERFORM P9000-DISPLAY-ERROR
                GOBACK
            END-IF

        END-IF.

    P200-MAINLINE.

        DISPLAY 'PROCESSING IN P200-MAINLINE'

*       CHECK STATUS CODE AND FIRST RECORD

        IF SW-END-OF-DB THEN
            DISPLAY 'NO RECORDS TO PROCESS!!'
        ELSE
            PERFORM UNTIL SW-END-OF-DB
                DISPLAY 'SUCCESSFUL READ :' IO-EMPLOYEE-RECORD
                MOVE EMPL-ID TO EMP-ID-VAL
                PERFORM P2000-GET-EMPPAY
                IF STATUS-CODE NOT EQUAL SPACES THEN
                    PERFORM P9000-DISPLAY-ERROR
                    GOBACK
                ELSE
                    MOVE PAY-EFF-DATE TO EFFDATE-VAL
                    SET SW-NOT-END-OF-EMPPAYHS TO TRUE
                    PERFORM P3000-GET-NEXT-EMPPAYHS
```

110

```
                UNTIL SW-END-OF-EMPPAYHS
         END-IF

         PERFORM P1000-GET-NEXT-ROOT
         IF STATUS-CODE = 'GB' THEN
             SET SW-END-OF-DB TO TRUE
             DISPLAY 'END OF DATABASE'
         END-IF

       END-PERFORM

    END-IF.

    DISPLAY 'FINISHED PROCESSING IN P200-MAINLINE'.

P300-TERMINATION.

    DISPLAY 'PROCESSING IN P300-TERMINATION'
    DISPLAY '** COBIMS9 - SUCCESSFULLY ENDED **'.

P1000-GET-NEXT-ROOT.

    DISPLAY '******************************'
    DISPLAY 'PROCESSING IN P1000-GET-NEXT-ROOT'.

    CALL 'CBLTDLI' USING FOUR,
         DLI-FUNCGN,
         PCB-MASK,
         IO-EMPLOYEE-RECORD,
         EMP-UNQUALIFIED-SSA.

P2000-GET-EMPPAY.

    DISPLAY 'PROCESSING IN P2000-GET-EMPPAY'.

    CALL 'CBLTDLI' USING FIVE,
         DLI-FUNCGNP,
         PCB-MASK,
         IO-EMPPAY-RECORD,
         EMP-QUALIFIED-SSA,
         EMPPAY-UNQUALIFIED-SSA.

P3000-GET-NEXT-EMPPAYHS.

    DISPLAY 'PROCESSING IN P3000-GET-NEXT-EMPPAYHS'.

    CALL 'CBLTDLI' USING SIX,
         DLI-FUNCGNP,
```

```
                    PCB-MASK,
                    IO-EMPPAYHS-RECORD,
                    EMP-QUALIFIED-SSA,
                    EMPPAY-QUALIFIED-SSA,
                    EMPPAYHS-UNQUALIFIED-SSA.

                    EVALUATE STATUS-CODE
                        WHEN ' '
                            DISPLAY 'GOOD READ OF EMPPAYHS : '
                                EMP-ID-VAL
                            MOVE PAY-ANN-PAY TO DIS-REG-PAY
                            MOVE PAY-AMT      TO DIS-SMT-PAY
                            DISPLAY 'PAY-DATE   : '  PAY-DATE
                            DISPLAY 'PAY-ANN-PAY: '  DIS-REG-PAY
                            DISPLAY 'PAY-AMT    : '  DIS-SMT-PAY
                        WHEN 'GE'
                        WHEN 'GB'
                            SET SW-END-OF-EMPPAYHS TO TRUE
                            DISPLAY 'NO MORE PAY HISTORY SEGMENTS'
                        WHEN OTHER
                            PERFORM P9000-DISPLAY-ERROR
                            SET SW-END-OF-EMPPAYHS TO TRUE
                            GOBACK
                    END-EVALUATE.

            P9000-DISPLAY-ERROR.

                DISPLAY 'ERROR ENCOUNTERED - DETAIL FOLLOWS'
                DISPLAY 'DBD-NAME1:'        DBD-NAME
                DISPLAY 'SEG-LEVEL1:'       SEG-LEVEL
                DISPLAY 'STATUS-CODE:'      STATUS-CODE
                DISPLAY 'PROC-OPT1 :'       PROC-OPT
                DISPLAY 'SEG-NAME1 :'       SEG-NAME
                DISPLAY 'KEY-FDBK1 :'       KEY-FDBK
                DISPLAY 'NUM-SENSEG1:'      NUM-SENSEG
                DISPLAY 'KEY-FDBK-AREA1:' KEY-FDBK-AREA.

        *    END OF SOURCE CODE
```

Compile, link, run. Here is the output.

```
** PROGRAM COBIMS9 START **
PROCESSING IN P100-INITIALIZATION
*********************************
PROCESSING IN P200-MAINLINE
SUCCESSFUL READ :1111 VEREEN                    CHARLES          12 201
PROCESSING IN P2000-GET-EMPPAY
PROCESSING IN P3000-GET-NEXT-EMPPAYHS
```

```
GOOD READ OF EMPPAYHS : 1111
PAY-DATE   : 20170115
PAY-ANN-PAY: 87000.00
PAY-AMT    : 3625.00
PROCESSING IN P3000-GET-NEXT-EMPPAYHS
GOOD READ OF EMPPAYHS : 1111
PAY-DATE   : 20170130
PAY-ANN-PAY: 87000.00
PAY-AMT    : 3625.00
PROCESSING IN P3000-GET-NEXT-EMPPAYHS
GOOD READ OF EMPPAYHS : 1111
PAY-DATE   : 20170215
PAY-ANN-PAY: 87000.00
PAY-AMT    : 3625.00
PROCESSING IN P3000-GET-NEXT-EMPPAYHS
GOOD READ OF EMPPAYHS : 1111
PAY-DATE   : 20170228
PAY-ANN-PAY: 87000.00
PAY-AMT    : 3625.00
PROCESSING IN P3000-GET-NEXT-EMPPAYHS
NO MORE PAY HISTORY SEGMENTS
*******************************
PROCESSING IN P1000-GET-NEXT-ROOT
SUCCESSFUL READ :1122 JENKINS              DEBORAH          05 201
PROCESSING IN P2000-GET-EMPPAY
PROCESSING IN P3000-GET-NEXT-EMPPAYHS
GOOD READ OF EMPPAYHS : 1122
PAY-DATE   : 20170115
PAY-ANN-PAY: 82000.00
PAY-AMT    : 3416.66
PROCESSING IN P3000-GET-NEXT-EMPPAYHS
GOOD READ OF EMPPAYHS : 1122
PAY-DATE   : 20170130
PAY-ANN-PAY: 82000.00
PAY-AMT    : 3416.66
PROCESSING IN P3000-GET-NEXT-EMPPAYHS
GOOD READ OF EMPPAYHS : 1122
PAY-DATE   : 20170215
PAY-ANN-PAY: 82000.00
PAY-AMT    : 3416.66
PROCESSING IN P3000-GET-NEXT-EMPPAYHS
GOOD READ OF EMPPAYHS : 1122
PAY-DATE   : 20170228
PAY-ANN-PAY: 82000.00
PAY-AMT    : 3416.66
PROCESSING IN P3000-GET-NEXT-EMPPAYHS
NO MORE PAY HISTORY SEGMENTS
*******************************
PROCESSING IN P1000-GET-NEXT-ROOT
SUCCESSFUL READ :3217 JOHNSON              EDWARD           04 201
PROCESSING IN P2000-GET-EMPPAY
PROCESSING IN P3000-GET-NEXT-EMPPAYHS
GOOD READ OF EMPPAYHS : 3217
PAY-DATE   : 20170115
```

```
PAY-ANN-PAY: 65000.00
PAY-AMT    : 2708.33
PROCESSING IN P3000-GET-NEXT-EMPPAYHS
GOOD READ OF EMPPAYHS : 3217
PAY-DATE   : 20170130
PAY-ANN-PAY: 65000.00
PAY-AMT    : 2708.33
PROCESSING IN P3000-GET-NEXT-EMPPAYHS
GOOD READ OF EMPPAYHS : 3217
PAY-DATE   : 20170215
PAY-ANN-PAY: 65000.00
PAY-AMT    : 2708.33
PROCESSING IN P3000-GET-NEXT-EMPPAYHS
GOOD READ OF EMPPAYHS : 3217
PAY-DATE   : 20170228
PAY-ANN-PAY: 65000.00
PAY-AMT    : 2708.33
PROCESSING IN P3000-GET-NEXT-EMPPAYHS
NO MORE PAY HISTORY SEGMENTS
********************************
PROCESSING IN P1000-GET-NEXT-ROOT
SUCCESSFUL READ :4175 TURNBULL                    FRED              01 201
PROCESSING IN P2000-GET-EMPPAY
PROCESSING IN P3000-GET-NEXT-EMPPAYHS
GOOD READ OF EMPPAYHS : 4175
PAY-DATE   : 20170115
PAY-ANN-PAY: 55000.00
PAY-AMT    : 2291.66
PROCESSING IN P3000-GET-NEXT-EMPPAYHS
GOOD READ OF EMPPAYHS : 4175
PAY-DATE   : 20170130
PAY-ANN-PAY: 55000.00
PAY-AMT    : 2291.66
PROCESSING IN P3000-GET-NEXT-EMPPAYHS
GOOD READ OF EMPPAYHS : 4175
PAY-DATE   : 20170215
PAY-ANN-PAY: 55000.00
PAY-AMT    : 2291.66
PROCESSING IN P3000-GET-NEXT-EMPPAYHS
GOOD READ OF EMPPAYHS : 4175
PAY-DATE   : 20170228
PAY-ANN-PAY: 55000.00
PAY-AMT    : 2291.66
PROCESSING IN P3000-GET-NEXT-EMPPAYHS
NO MORE PAY HISTORY SEGMENTS
******************************
PROCESSING IN P1000-GET-NEXT-ROOT
SUCCESSFUL READ :4720 SCHULTZ                      TIM               09 201
PROCESSING IN P2000-GET-EMPPAY
PROCESSING IN P3000-GET-NEXT-EMPPAYHS
GOOD READ OF EMPPAYHS : 4720
PAY-DATE   : 20170115
PAY-ANN-PAY: 80000.00
PAY-AMT    : 3333.33
```

```
PROCESSING IN P3000-GET-NEXT-EMPPAYHS
GOOD READ OF EMPPAYHS : 4720
PAY-DATE   : 20170130
PAY-ANN-PAY: 80000.00
PAY-AMT    : 3333.33
PROCESSING IN P3000-GET-NEXT-EMPPAYHS
GOOD READ OF EMPPAYHS : 4720
PAY-DATE   : 20170215
PAY-ANN-PAY: 80000.00
PAY-AMT    : 3333.33
PROCESSING IN P3000-GET-NEXT-EMPPAYHS
GOOD READ OF EMPPAYHS : 4720
PAY-DATE   : 20170228
PAY-ANN-PAY: 80000.00
PAY-AMT    : 3333.33
PROCESSING IN P3000-GET-NEXT-EMPPAYHS
NO MORE PAY HISTORY SEGMENTS
********************************
PROCESSING IN P1000-GET-NEXT-ROOT
SUCCESSFUL READ :4836 SMITH                  SANDRA           03 201
PROCESSING IN P2000-GET-EMPPAY
PROCESSING IN P3000-GET-NEXT-EMPPAYHS
GOOD READ OF EMPPAYHS : 4836
PAY-DATE   : 20170115
PAY-ANN-PAY: 62000.00
PAY-AMT    : 2583.33
PROCESSING IN P3000-GET-NEXT-EMPPAYHS
GOOD READ OF EMPPAYHS : 4836
PAY-DATE   : 20170130
PAY-ANN-PAY: 62000.00
PAY-AMT    : 2583.33
PROCESSING IN P3000-GET-NEXT-EMPPAYHS
GOOD READ OF EMPPAYHS : 4836
PAY-DATE   : 20170215
PAY-ANN-PAY: 62000.00
PAY-AMT    : 2583.33
PROCESSING IN P3000-GET-NEXT-EMPPAYHS
GOOD READ OF EMPPAYHS : 4836
PAY-DATE   : 20170228
PAY-ANN-PAY: 62000.00
PAY-AMT    : 2583.33
PROCESSING IN P3000-GET-NEXT-EMPPAYHS
NO MORE PAY HISTORY SEGMENTS
********************************
PROCESSING IN P1000-GET-NEXT-ROOT
SUCCESSFUL READ :6288 WILLARD                JOE              06 201
PROCESSING IN P2000-GET-EMPPAY
PROCESSING IN P3000-GET-NEXT-EMPPAYHS
GOOD READ OF EMPPAYHS : 6288
PAY-DATE   : 20170115
PAY-ANN-PAY: 70000.00
PAY-AMT    : 2916.66
PROCESSING IN P3000-GET-NEXT-EMPPAYHS
GOOD READ OF EMPPAYHS : 6288
```

```
PAY-DATE   : 20170130
PAY-ANN-PAY: 70000.00
PAY-AMT    :  2916.66
PROCESSING IN P3000-GET-NEXT-EMPPAYHS
GOOD READ OF EMPPAYHS : 6288
PAY-DATE   : 20170215
PAY-ANN-PAY: 70000.00
PAY-AMT    :  2916.66
PROCESSING IN P3000-GET-NEXT-EMPPAYHS
GOOD READ OF EMPPAYHS : 6288
PAY-DATE   : 20170228
PAY-ANN-PAY: 70000.00
PAY-AMT    :  2916.66
PROCESSING IN P3000-GET-NEXT-EMPPAYHS
NO MORE PAY HISTORY SEGMENTS
* * * * * * * * * * * * * * * * * * * * * * * * * * * * * * * *
PROCESSING IN P1000-GET-NEXT-ROOT
SUCCESSFUL READ :7459 STEWART             BETTY              07 201
PROCESSING IN P2000-GET-EMPPAY
PROCESSING IN P3000-GET-NEXT-EMPPAYHS
GOOD READ OF EMPPAYHS : 7459
PAY-DATE   : 20170115
PAY-ANN-PAY: 85000.00
PAY-AMT    :  3541.66
PROCESSING IN P3000-GET-NEXT-EMPPAYHS
GOOD READ OF EMPPAYHS : 7459
PAY-DATE   : 20170130
PAY-ANN-PAY: 85000.00
PAY-AMT    :  3541.66
PROCESSING IN P3000-GET-NEXT-EMPPAYHS
GOOD READ OF EMPPAYHS : 7459
PAY-DATE   : 20170215
PAY-ANN-PAY: 85000.00
PAY-AMT    :  3541.66
PROCESSING IN P3000-GET-NEXT-EMPPAYHS
GOOD READ OF EMPPAYHS : 7459
PAY-DATE   : 20170228
PAY-ANN-PAY: 85000.00
PAY-AMT    :  3541.66
PROCESSING IN P3000-GET-NEXT-EMPPAYHS
NO MORE PAY HISTORY SEGMENTS
* * * * * * * * * * * * * * * * * * * * * * * * * * * * * * * *
PROCESSING IN P1000-GET-NEXT-ROOT
END OF DATABASE
FINISHED PROCESSING IN P200-MAINLINE
PROCESSING IN P300-TERMINATION
** COBIMS9 - SUCCESSFULLY ENDED **
```

Ok I think we've covered the root-child relationships enough. You have some models
to use for most anything you'd want to do in the hierarchy. Time to move on to other
topics. First, here's the PLI code that corresponds to the COBIMS9 program.

```
PLIIMS9: PROCEDURE (DB_PTR_PCB) OPTIONS(MAIN);
/*********************************************************************
* PROGRAM NAME :   PLIIMS9 -  READ EMPLOYEE HISTORY PAY RECORDS     *
*                  UNDER THEIR EMPPAY PARENTS. QUALIFIED SSA'S       *
*                  MUST BE PROVIDED FOR BOTH THE EMPLOYEE AND        *
*                  EMPPAY SEGMENTS.                                  *
*********************************************************************/
/*********************************************************************
/*                W O R K I N G   S T O R A G E                     *
*********************************************************************/

   DCL SW_END_OF_DB            STATIC BIT(01) INIT('0'B);
   DCL SW_END_OF_EMPPAYHS      STATIC BIT(01) INIT('0'B);
   DCL ONCODE                  BUILTIN;
   DCL DB_PTR_PCB              POINTER;

   DCL PLITDLI                 EXTERNAL ENTRY;

   DCL 01 DLI_FUNCTIONS,
         05 DLI_FUNCISRT       CHAR(04) INIT ('ISRT'),
         05 DLI_FUNCGU         CHAR(04) INIT ('GU  '),
         05 DLI_FUNCGN         CHAR(04) INIT ('GN  '),
         05 DLI_FUNCGHU        CHAR(04) INIT ('GHU '),
         05 DLI_FUNCGNP        CHAR(04) INIT ('GNP '),
         05 DLI_FUNCREPL       CHAR(04) INIT ('REPL'),
         05 DLI_FUNCDLET       CHAR(04) INIT ('DLET'),
         05 DLI_FUNCXRST       CHAR(04) INIT ('XRST'),
         05 DLI_FUNCCHKP       CHAR(04) INIT ('CHKP'),
         05 DLI_FUNCROLL       CHAR(04) INIT ('ROLL');

   DCL 01 IO_EMPLOYEE_RECORD,
         05  EMPL_ID           CHAR(04),
         05  FILLER1           CHAR(01),
         05  EMPL_LNAME        CHAR(30),
         05  FILLER2           CHAR(01),
         05  EMPL_FNAME        CHAR(20),
         05  FILLER3           CHAR(01),
         05  EMPL_YRS_SRV      CHAR(02),
         05  FILLER4           CHAR(01),
         05  EMPL_PRM_DTE      CHAR(10),
         05  FILLER5           CHAR(10);

   DCL 01 IO_EMPPAY_RECORD,
         05  PAY_EFF_DATE      CHAR(8),
         05  PAY_REG_PAY       FIXED DEC (8,2),
         05  PAY_BON_PAY       FIXED DEC (8,2),
         05  SEMIMTH_PAY       FIXED DEC (8,2),
         05  FILLER55          CHAR(57);

   DCL 01 IO_EMPPAYHS_RECORD,
         05  PAY_DATE          CHAR(8),
         05  PAY_ANN_PAY       FIXED DEC (8,2),
         05  PAY_AMT           FIXED DEC (8,2),
         05  FILLER65          CHAR(62);
```

```
DCL 01 PCB_MASK              BASED(DB_PTR_PCB),
        05  DBD_NAME         CHAR(08),
        05  SEG_LEVEL        CHAR(02),
        05  STATUS_CODE      CHAR(02),
        05  PROC_OPT         CHAR(04),
        05  FILLER6          FIXED BIN (31),
        05  SEG_NAME         CHAR(08),
        05  KEY_FDBK         FIXED BIN (31),
        05  NUM_SENSEG       FIXED BIN (31),
        05  KEY_FDBK_AREA,
        10  EMPLOYEE_ID      CHAR(04),
        10  EMP_PAY_DATE     CHAR(08);

DCL 01 EMP_UNQUALIFIED_SSA,
        05  SEGNAME          CHAR(08) INIT ('EMPLOYEE'),
        05  FILLER7          CHAR(01) INIT (' ');

DCL 01 EMP_QUALIFIED_SSA,
        05  SEGNAME          CHAR(08) INIT('EMPLOYEE'),
        05  FILLER8          CHAR(01) INIT('('),
        05  FIELD            CHAR(08) INIT('EMPID'),
        05  OPER             CHAR(02) INIT(' ='),
        05  EMP_ID_VAL       CHAR(04) INIT('    '),
        05  FILLER9          CHAR(01) INIT(')');

DCL 01 EMPPAY_UNQUALIFIED_SSA,
        05  SEGNAME          CHAR(08) INIT('EMPPAY  '),
        05  FILLER10         CHAR(01) INIT(' ');

DCL 01 EMPPAY_QUALIFIED_SSA,
        05  SEGNAME          CHAR(08) INIT('EMPPAY  '),
        05  FILLER11         CHAR(01) INIT('('),
        05  FIELD            CHAR(08) INIT('EFFDATE '),
        05  OPER             CHAR(02) INIT(' ='),
        05  EFFDATE_VAL      CHAR(08) INIT('        '),
        05  FILLER12         CHAR(01) INIT(')');

DCL 01 EMPPAYHS_UNQUALIFIED_SSA,
        05  SEGNAME          CHAR(08) INIT('EMPPAYHS'),
        05  FILLER13         CHAR(01) INIT(' ');

DCL WS_PAY_DATE             CHAR(08) INIT ('20170215');

DCL SEG_IO_AREA             CHAR(80) INIT (' ');

DCL THREE                   FIXED BIN (31) INIT(3);
DCL FOUR                    FIXED BIN (31) INIT(4);
DCL FIVE                    FIXED BIN (31) INIT(5);
DCL SIX                     FIXED BIN (31) INIT(6);

/*******************************************************************
/*            P R O G R A M   M A I N L I N E             *
*******************************************************************/
```

```
CALL P100_INITIALIZATION;
CALL P200_MAINLINE;
CALL P300_TERMINATION;

P100_INITIALIZATION: PROC;

    PUT SKIP LIST ('PLIIMS9: READ EMPPAYHS RECORDS');
    PUT SKIP LIST ('PROCESSING IN P100_INITIALIZATION');

    IO_EMPLOYEE_RECORD  = '';
    IO_EMPPAY_RECORD  = '';
    IO_EMPPAYHS_RECORD  = '';
    PCB_MASK = '';

    /* DO INITIAL DB READ FOR FIRST EMPLOYEE RECORD */

    CALL PLITDLI (FOUR,
                  DLI_FUNCGN,
                  PCB_MASK,
                  IO_EMPLOYEE_RECORD,
                  EMP_UNQUALIFIED_SSA);

    IF STATUS_CODE = '  ' THEN;
    ELSE
        IF STATUS_CODE = 'GB' THEN
            DO;
                SW_END_OF_DB = '1'B;
                PUT SKIP LIST ('** END OF DATABASE');
            END;
        ELSE
            DO;
                CALL P9000_DISPLAY_ERROR;
                RETURN;
            END;

END P100_INITIALIZATION;

P200_MAINLINE: PROC;

    /*  MAIN LOOP - WALK THROUGH THE DATABASE GETTING EMPLOYEE
                    PAY HISTORY SEGMENTS.                    */

    IF SW_END_OF_DB THEN
        PUT SKIP LIST ('NO RECORDS TO PROCESS!!');
    ELSE
        DO WHILE (¬SW_END_OF_DB);

            PUT SKIP LIST ('SUCCESSFUL READ OF EMPLOYEE ' || EMPL_ID);

            /* ASSIGN KEY FOR EMPLOYEE LEVEL SSA AND GET EMPPAY SEG */

            EMP_ID_VAL = EMPL_ID;
            CALL P2000_GET_EMPPAY;
            IF STATUS_CODE ¬= '  ' THEN
```

```
                    DO;
                        PUT SKIP LIST ('ERROR READING EMPPAY');
                        CALL P9000_DISPLAY_ERROR;
                        RETURN;
                    END;
                ELSE
                    DO;
                        PUT SKIP LIST ('SUCCESSFUL PAY READ : '
                            || IO_EMPPAY_RECORD);

                        /* ASSIGN KEY FOR EMPPAY LEVEL SSA FOR SEG */

                        EFFDATE_VAL = PAY_EFF_DATE;

                        SW_END_OF_EMPPAYHS = '0'B;
                        DO WHILE (¬SW_END_OF_EMPPAYHS);
                            CALL P3000_GET_NEXT_EMPPAYHS;
                        END; /* DO WHILE */

                        CALL P1000_GET_NEXT_ROOT;
                        IF STATUS_CODE = 'GB' THEN
                            DO;
                                SW_END_OF_DB = '1'B;
                                PUT SKIP LIST ('** END OF DATABASE');
                            END;
                        ELSE
                            IF STATUS_CODE = '  ' THEN;
                            ELSE
                                DO;
                                    CALL P9000_DISPLAY_ERROR;
                                    RETURN;
                                END;

                    END; /* SUCCESSFULLY RETRIEVED PAY SEG */

            END; /* DO WHILE */

    PUT SKIP LIST ('FINISHED PROCESSING IN P200_MAINLINE');

END P200_MAINLINE;

P300_TERMINATION: PROC;

    PUT SKIP LIST ('PLIIMS9 - SUCCESSFULLY ENDED');

END P300_TERMINATION;

P1000_GET_NEXT_ROOT: PROC;

    PUT SKIP LIST ('PROCESSING IN P1000_GET_NEXT_ROOT');

    CALL PLITDLI (FOUR,
                  DLI_FUNCGN,
                  PCB_MASK,
```

```
                    IO_EMPLOYEE_RECORD,
                    EMP_UNQUALIFIED_SSA);

END P1000_GET_NEXT_ROOT;

P2000_GET_EMPPAY: PROC;

   PUT SKIP LIST ('PROCESSING IN P2000_GET_EMPPAY');

   CALL PLITDLI (FIVE,
                 DLI_FUNCGNP,
                 PCB_MASK,
                 IO_EMPPAY_RECORD,
                 EMP_QUALIFIED_SSA,
                 EMPPAY_UNQUALIFIED_SSA);

END P2000_GET_EMPPAY;

P3000_GET_NEXT_EMPPAYHS: PROC;

   PUT SKIP LIST ('PROCESSING IN P3000_GET_NEXT_EMPPAYHS');

   CALL PLITDLI (SIX,
                 DLI_FUNCGNP,
                 PCB_MASK,
                 IO_EMPPAYHS_RECORD,
                 EMP_QUALIFIED_SSA,
                 EMPPAY_QUALIFIED_SSA,
                 EMPPAYHS_UNQUALIFIED_SSA);

   SELECT (STATUS_CODE);
      WHEN ('  ')
         DO;
            PUT SKIP LIST ('GOOD READ OF EMPPAYHS : '
               || EMP_ID_VAL);
            PUT SKIP LIST ('PAY_DATE    : ' || PAY_DATE);
            PUT SKIP LIST ('PAY_ANN_PAY: ' || PAY_ANN_PAY);
            PUT SKIP LIST ('PAY_AMT     : ' || PAY_AMT);
         END;
      WHEN ('GE','GB')
         DO;
            SW_END_OF_EMPPAYHS = '1'B;
            PUT SKIP LIST ('NO MORE PAY HISTORY SEGMENTS');
         END;
      OTHERWISE
         DO;
            CALL P9000_DISPLAY_ERROR;
            SW_END_OF_EMPPAYHS = '1'B;
         END;

   END; /* SELECT */

END P3000_GET_NEXT_EMPPAYHS;
```

```
P9000_DISPLAY_ERROR: PROC;

    PUT SKIP LIST ('ERROR ENCOUNTERED - DETAIL FOLLOWS');
    PUT SKIP DATA (IO_EMPPAY_RECORD);
    PUT SKIP LIST ('DBD_NAME1:'  ||  DBD_NAME);
    PUT SKIP LIST ('SEG_LEVEL1:' || SEG_LEVEL);
    PUT SKIP LIST ('STATUS_CODE:' || STATUS_CODE);
    PUT SKIP LIST ('PROC_OPT1 :'  || PROC_OPT);
    PUT SKIP LIST ('SEG_NAME1 :'  || SEG_NAME);
    PUT SKIP LIST ('KEY_FDBK1 :'  || KEY_FDBK);
    PUT SKIP LIST ('NUM_SENSEG1:' || NUM_SENSEG);
    PUT SKIP LIST ('KEY_FDBK_AREA1:' || KEY_FDBK_AREA);

END P9000_DISPLAY_ERROR;

END PLIIMS9;
```

Additional IMS Programming Features

Retrieve Segments Using Searchable Fields

So far all the qualified SSA retrievals we've done have been based on a segment **key**. It is also possible to retrieve IMS segments by a searchable field that is not the key. For this example with program COBIMSA we will create a new field for our EMPLOYEE record layout, and then define this field in our DBD. Then we will write a program to search based on the new EMPSSN field which is the employee social security number.

Ok, where shall we put the field? We have a 9 byte social security number field, and we have 10 bytes of filler at the end of the record. Let's use the last 9 bytes of the record. Here is our new layout.

```
01 IO-EMPLOYEE-RECORD.
   05   FILLER        PIC X(06).
   05   EMP-ID        PIC X(04).
   05   FILLER        PIC X(01).
   05   EMPL-LNAME    PIC X(30).
   05   FILLER        PIC X(01).
   05   EMPL-FNAME    PIC X(20).
   05   FILLER        PIC X(01).
   05   EMPL-YRS-SRV  PIC X(02).
   05   FILLER        PIC X(01).
   05   EMPL-PRM-DTE  PIC X(10).
   05   FILLER        PIC X(01).
   05   EMPL-SSN      PIC X(09).
```

Now let's assign EMPL-SSN values to the original flat file we used to load the database. Here it is:

```
BROWSE    USER01.EMPIFILE                      Line 00000000 Col 001 080
----+----1----+----2----+----3----+----4----+----5----+----6----+----7----+----8
 Command ===>                                            Scroll ===> CSR
****************************** Top of Data ********************************
1111 VEREEN                 CHARLES             12 2017-01-01 937253058
1122 JENKINS                DEBORAH             05 2017-01-01 435092366
3217 JOHNSON                EDWARD              04 2017-01-01 397342007
4175 TURNBULL               FRED                01 2016-12-01 542083017
4720 SCHULTZ                TIM                 09 2017-01-01 650450254
4836 SMITH                  SANDRA              03 2017-01-01 028374669
6288 WILLARD                JOE                 06 2016-01-01 209883920
7459 STEWART                BETTY               07 2016-07-31 019572830
9134 FRANKLIN               BRIANNA             00 2016-10-01 937293598
***************************** Bottom of Data *****************************
```

Now let's delete all existing records in the database (you can use File Manager for this as explained earlier in the chapter). Then let's run COBIMS1 to reload the database from our flat file which now includes the EMPL-SSN values. Now we can browse the database and verify that the EMPL-SSN field is populated (you will need to scroll to the right to see the EMPSSN field).

```
Browse              USER01.IMS.EMPLOYEE.CLUSTER                Top of 9
Command ===>                                                   Scroll PAGE
                        Type KSDS     RBA                      Format CHAR
Key                                         Col 10
>----+----20---+----3----+----4----+----5----+----6----+----7----+----8----+---
****  Top of data  ****
1 VEREEN                     CHARLES            12 2017-01-01 937253058..
2 JENKINS                    DEBORAH            05 2017-01-01 435092366..
7 JOHNSON                    EDWARD             04 2017-01-01 397342007..
5 TURNBULL                   FRED               01 2016-12-01 542083017..
0 SCHULTZ                    TIM                09 2017-01-01 650450254..
6 SMITH                      SANDRA             03 2017-01-01 028374669..
8 WILLARD                    JOE                06 2016-01-01 209883920..
9 STEWART                    BETTY              07 2016-07-31 019572830..
4 FRANKLIN                   BRIANNA            00 2016-10-01 937293598..
```

Ok, next step. To be able to search on a field in an IMS segment, the field must be defined in the DBD. Recall our original code for the DBD is as follows:

```
PRINT NOGEN
DBD NAME=EMPLOYEE,ACCESS=HISAM
DATASET DD1=EMPLOYEE,OVFLW=EMPLFLW
SEGM NAME=EMPLOYEE,PARENT=0,BYTES=80
FIELD NAME=(EMPID,SEQ,U),BYTES=04,START=1,TYPE=C
SEGM NAME=EMPPAY,PARENT=EMPLOYEE,BYTES=23
FIELD NAME=(EFFDATE,SEQ,U),START=1,BYTES=8,TYPE=C
SEGM  NAME=EMPPAYHS,PARENT=EMPPAY,BYTES=18
FIELD NAME=(PAYDATE,SEQ,U),START=1,BYTES=8,TYPE=C
DBDGEN
FINISH
END
```

The only searchable field right now on the EMPLOYEE segment is the primary key EM-PID. To make the EMPSSN field searchable we must add it to the DBD. The appropriate code is bolded below. Note that EMPSSN starts in position 72 of the record and is 9 bytes in length.

```
PRINT NOGEN
DBD NAME=EMPLOYEE,ACCESS=HISAM
```

```
DATASET DD1=EMPLOYEE,OVFLW=EMPLFLW
SEGM NAME=EMPLOYEE,PARENT=0,BYTES=80
FIELD NAME=(EMPID,SEQ,U),BYTES=04,START=1,TYPE=C
FIELD NAME=EMPSSN,START=72,BYTES=9,TYPE=C
SEGM NAME=EMPPAY,PARENT=EMPLOYEE,BYTES=23
FIELD NAME=(EFFDATE,SEQ,U),START=1,BYTES=8,TYPE=C
SEGM  NAME=EMPPAYHS,PARENT=EMPPAY,BYTES=18
FIELD NAME=(PAYDATE,SEQ,U),START=1,BYTES=8,TYPE=C
DBDGEN
FINISH
END
```

Go ahead and run the DBD gen process.

Next we can write a program to search on the EMPSSN field. We can clone the COBIMS2 program to make COBIMSA. One change we must make is to use a different qualified SSA than the one we started with. We need only change the field name in the SSA and create a value field with an appropriate specification (in this case a 9 position character field for the SSN key).

Here is our new structure:

```
01 EMP-QUALIFIED-SSA-EMPSSN.
   05  SEGNAME     PIC X(08) VALUE 'EMPLOYEE'.
   05  FILLER      PIC X(01) VALUE '('.
   05  FIELD       PIC X(08) VALUE 'EMPSSN'.
   05  OPER        PIC X(02) VALUE ' ='.
   05  EMPSSN-VAL  PIC X(09) VALUE '         '.
   05  FILLER      PIC X(01) VALUE ')'.
```

Naturally you must load the EMPSSN-VAL variable with the value you are looking for. Let's use the social security number 937253058 for Charles Vereen who is employee number 1111. Here is our COBOL program source.

```
 IDENTIFICATION DIVISION.
 PROGRAM-ID. COBIMSA.
 ***************************************************
 *   RETRIEVE A RECORD FROM IMS EMPLOYEE DATABASE   *
 *   USING SEARCHABLE FIELD EMPSSN                  *
 ***************************************************

 ENVIRONMENT DIVISION.
 DATA DIVISION.
```

```
**************************************************
*  W O R K I N G   S T O R A G E   S E C T I O N   *
**************************************************

 WORKING-STORAGE SECTION.

 01 SEG-IO-AREA     PIC X(80).

 01 DLI-FUNCTIONS.
    05 DLI-FUNCISRT  PIC X(4) VALUE 'ISRT'.
    05 DLI-FUNCGU    PIC X(4) VALUE 'GU  '.
    05 DLI-FUNCGN    PIC X(4) VALUE 'GN  '.
    05 DLI-FUNCGHU   PIC X(4) VALUE 'GHU '.
    05 DLI-FUNCGNP   PIC X(4) VALUE 'GNP '.
    05 DLI-FUNCREPL  PIC X(4) VALUE 'REPL'.
    05 DLI-FUNCDLET  PIC X(4) VALUE 'DLET'.
    05 DLI-FUNCXRST  PIC X(4) VALUE 'XRST'.
    05 DLI-FUNCCKPT  PIC X(4) VALUE 'CKPT'.

 01 EMP-UNQUALIFIED-SSA.
    05  SEGNAME     PIC X(08) VALUE 'EMPLOYEE'.
    05  FILLER      PIC X(01) VALUE ' '.

 01 EMP-QUALIFIED-SSA.
    05  SEGNAME     PIC X(08) VALUE 'EMPLOYEE'.
    05  FILLER      PIC X(01) VALUE '('.
    05  FIELD       PIC X(08) VALUE 'EMPID'.
    05  OPER        PIC X(02) VALUE ' ='.
    05  EMP-ID-VAL  PIC X(04) VALUE '    '.
    05  FILLER      PIC X(01) VALUE ')'.

 01 EMP-QUALIFIED-SSA-EMPSSN.
    05  SEGNAME     PIC X(08) VALUE 'EMPLOYEE'.
    05  FILLER      PIC X(01) VALUE '('.
    05  FIELD       PIC X(08) VALUE 'EMPSSN'.
    05  OPER        PIC X(02) VALUE ' ='.
    05  EMPSSN-VAL  PIC X(09) VALUE '         '.
    05  FILLER      PIC X(01) VALUE ')'.

 01 IMS-RET-CODES.
    05 THREE        PIC S9(9) COMP VALUE +3.
    05 FOUR         PIC S9(9) COMP VALUE +4.
    05 FIVE         PIC S9(9) COMP VALUE +5.
    05 SIX          PIC S9(9) COMP VALUE +6.

 LINKAGE SECTION.
 01 PCB-MASK.
    03 DBD-NAME        PIC X(8).
```

```cobol
   03 SEG-LEVEL       PIC XX.
   03 STATUS-CODE     PIC XX.
   03 PROC-OPT        PIC X(4).
   03 FILLER          PIC X(4).
   03 SEG-NAME        PIC X(8).
   03 KEY-FDBK        PIC S9(5) COMP.
   03 NUM-SENSEG      PIC S9(5) COMP.
   03 KEY-FDBK-AREA.
      05 EMPLOYEE-ID  PIC X(04).
      05 EMPPAYHS     PIC X(08).

PROCEDURE DIVISION.

   INITIALIZE PCB-MASK
   ENTRY 'DLITCBL' USING PCB-MASK

   PERFORM P100-INITIALIZATION.
   PERFORM P200-MAINLINE.
   PERFORM P300-TERMINATION.
   GOBACK.


P100-INITIALIZATION.

   DISPLAY '** PROGRAM COBIMSA START **'
   DISPLAY 'PROCESSING IN P100-INITIALIZATION'.


P200-MAINLINE.

   DISPLAY 'PROCESSING IN P200-MAINLINE'

   MOVE '937253058' TO EMPSSN-VAL

   DISPLAY 'EMP-QUALIFIED-SSA-EMPSSN '
      EMP-QUALIFIED-SSA-EMPSSN

   CALL 'CBLTDLI' USING FOUR,
                  DLI-FUNCGU,
                  PCB-MASK,
                  SEG-IO-AREA,
                  EMP-QUALIFIED-SSA-EMPSSN

   IF STATUS-CODE = '  '
      DISPLAY 'SUCCESSFUL GET CALL  '
      DISPLAY 'SEG-IO-ARE : ' SEG-IO-AREA
   ELSE
      DISPLAY 'ERROR IN FETCH :' STATUS-CODE
      PERFORM P400-DISPLAY-ERROR
   END-IF.
```

```
    P300-TERMINATION.

        DISPLAY 'PROCESSING IN P300-TERMINATION'
        DISPLAY '** COBIMSA - SUCCESSFULLY ENDED **'.

    P400-DISPLAY-ERROR.

        DISPLAY 'ERROR ENCOUNTERED - DETAIL FOLLOWS'
        DISPLAY 'SEG-IO-AREA      :' SEG-IO-AREA
        DISPLAY 'DBD-NAME1:'      DBD-NAME
        DISPLAY 'SEG-LEVEL1:'     SEG-LEVEL
        DISPLAY 'STATUS-CODE:'    STATUS-CODE
        DISPLAY 'PROC-OPT1 :'     PROC-OPT
        DISPLAY 'SEG-NAME1 :'     SEG-NAME
        DISPLAY 'KEY-FDBK1 :'     KEY-FDBK
        DISPLAY 'NUM-SENSEG1:'    NUM-SENSEG
        DISPLAY 'KEY-FDBK-AREA1:' KEY-FDBK-AREA.

    *   END OF SOURCE CODE
```

Again we compile, link and execute. Here's the output:

```
** PROGRAM COBIMSA START **
PROCESSING IN P100-INITIALIZATION
PROCESSING IN P200-MAINLINE
EMP-QUALIFIED-SSA-EMPSSN EMPLOYEE(EMPSSN   =937253058)
SUCCESSFUL GET CALL
SEG-IO-ARE : 1111 VEREEN          CHARLES              12 2017-01-01 937253058
PROCESSING IN P300-TERMINATION
** COBIMSA - SUCCESSFULLY ENDED **
```

As you can see, we retrieved the desired record using the EMPSSN search field. So keep in mind that you can search on fields other than the key field as long as they are defined in the DBD. If you are going to be searching on a non-indexed field often, you'll want to check with your DBA about possibly defining a secondary index.

Here's the PLI code for this example.

```
PLIIMSA: PROCEDURE (DB_PTR_PCB) OPTIONS(MAIN);
/****************************************************************
* PROGRAM NAME :   PLIIMSA - RETRIEVE A RECORD FROM EMPLOYEE DB   *
*                          USING SEARCH FIELD EMPSSN.            *
****************************************************************/

/****************************************************************
```

128

```
/*              W O R K I N G   S T O R A G E                    *
*********************************************************************/

    DCL ONCODE                  BUILTIN;
    DCL DB_PTR_PCB              POINTER;
    DCL PLITDLI                 EXTERNAL ENTRY;

    DCL 01 DLI_FUNCTIONS,
           05 DLI_FUNCISRT      CHAR(04) INIT ('ISRT'),
           05 DLI_FUNCGU        CHAR(04) INIT ('GU  '),
           05 DLI_FUNCGN        CHAR(04) INIT ('GN  '),
           05 DLI_FUNCGHU       CHAR(04) INIT ('GHU '),
           05 DLI_FUNCGNP       CHAR(04) INIT ('GNP '),
           05 DLI_FUNCREPL      CHAR(04) INIT ('REPL'),
           05 DLI_FUNCDLET      CHAR(04) INIT ('DLET'),
           05 DLI_FUNCXRST      CHAR(04) INIT ('XRST'),
           05 DLI_FUNCCHKP      CHAR(04) INIT ('CHKP'),
           05 DLI_FUNCROLL      CHAR(04) INIT ('ROLL');

    DCL 01 IO_EMPLOYEE_RECORD,
           05  EMPL_ID_IN       CHAR(04),
           05  FILLER1          CHAR(01),
           05  EMPL_LNAME       CHAR(30),
           05  FILLER2          CHAR(01),
           05  EMPL_FNAME       CHAR(20),
           05  FILLER3          CHAR(01),
           05  EMPL_YRS_SRV     CHAR(02),
           05  FILLER4          CHAR(01),
           05  EMPL_PRM_DTE     CHAR(10),
           05  FILLER5          CHAR(01),
           05  EMPL_SSN         CHAR(09);

    DCL 01 PCB_MASK             BASED(DB_PTR_PCB),
           05 DBD_NAME          CHAR(08),
           05 SEG_LEVEL         CHAR(02),
           05 STATUS_CODE       CHAR(02),
           05 PROC_OPT          CHAR(04),
           05 FILLER6           FIXED BIN (31),
           05 SEG_NAME          CHAR(08),
           05 KEY_FDBK          FIXED BIN (31),
           05 NUM_SENSEG        FIXED BIN (31),
           05 KEY_FDBK_AREA,
              10 EMPLOYEE_ID    CHAR(04);

    DCL 01 EMP_UNQUALIFIED_SSA,
           05  SEGNAME          CHAR(08) INIT ('EMPLOYEE'),
           05  FILLER7          CHAR(01) INIT (' ');

    DCL 01 EMP_QUALIFIED_SSA,
           05  SEGNAME          CHAR(08) INIT('EMPLOYEE'),
           05  FILLER8          CHAR(01) INIT('('),
           05  FIELD            CHAR(08) INIT('EMPID'),
           05  OPER             CHAR(02) INIT(' ='),
           05  EMP_ID_VAL       CHAR(04) INIT('    '),
```

```
           05   FILLER9             CHAR(01) INIT(')');

     DCL 01 EMP_QUALIFIED_SSA_EMPSSN,
           05   SEGNAME             CHAR(08) INIT('EMPLOYEE'),
           05   FILLER10            CHAR(01) INIT('('),
           05   FIELD               CHAR(08) INIT('EMPSSN'),
           05   OPER                CHAR(02) INIT(' ='),
           05   EMPSSN_VAL          CHAR(09) INIT('         '),
           05   FILLER11            CHAR(01) INIT(')');

     DCL THREE                      FIXED BIN (31) INIT(3);
     DCL FOUR                       FIXED BIN (31) INIT(4);
     DCL FIVE                       FIXED BIN (31) INIT(5);
     DCL SIX                        FIXED BIN (31) INIT(6);

/**********************************************************************
/*                 P R O G R A M   M A I N L I N E               *
**********************************************************************/

CALL P100_INITIALIZATION;
CALL P200_MAINLINE;
CALL P300_TERMINATION;

P100_INITIALIZATION: PROC;

    PUT SKIP LIST ('PLIIMSA: GET RECORD FROM EMPLOYEE DB USING EMPSS');
    IO_EMPLOYEE_RECORD  = '';

END P100_INITIALIZATION;

P200_MAINLINE: PROC;

   /*  SET THE EMPLOYEE SEGMENT SEARCH ARGUMENT AND CALL PLITDLI */

    EMPSSN_VAL = '937253058';

    CALL PLITDLI (FOUR,
                  DLI_FUNCGU,
                  PCB_MASK,
                  IO_EMPLOYEE_RECORD,
                  EMP_QUALIFIED_SSA_EMPSSN);

   IF STATUS_CODE = '  ' THEN
      DO;
         PUT SKIP LIST ('SUCCESSFUL RETRIEVAL - SSN: ' || EMPL_SSN);
         PUT SKIP DATA(IO_EMPLOYEE_RECORD);
      END;
   ELSE
      CALL P400_DISPLAY_ERROR;

END P200_MAINLINE;

P300_TERMINATION: PROC;
```

130

```
        PUT SKIP LIST ('PLIIMSA - ENDED SUCCESSFULLY');

END P300_TERMINATION;

P400_DISPLAY_ERROR: PROC;

        PUT SKIP LIST ('ERROR ENCOUNTERED - DETAIL FOLLOWS');
        PUT SKIP LIST ('SEG_IO_AREA      :' || SEG_IO_AREA);
        PUT SKIP LIST ('DBD_NAME1:' || DBD_NAME);
        PUT SKIP LIST ('SEG_LEVEL1:' || SEG_LEVEL);
        PUT SKIP LIST ('STATUS_CODE:' || STATUS_CODE);
        PUT SKIP LIST ('PROC_OPT1 :' || PROC_OPT);
        PUT SKIP LIST ('SEG_NAME1 :' || SEG_NAME);
        PUT SKIP LIST ('KEY_FDBK1 :' || KEY_FDBK);
        PUT SKIP LIST ('NUM_SENSEG1:' || NUM_SENSEG);
        PUT SKIP LIST ('KEY_FDBK_AREA1:' || KEY_FDBK_AREA);

END P400_DISPLAY_ERROR;

END PLIIMSA;
```

Retrieve Segments Using Boolean SSAs

The qualified SSA retrievals we've done so far have searched using a field value that is equal to a single searchable field. It is also possible to retrieve IMS segments using other Boolean operators such as greater than or less than. Additionally, you can specify more than one operator, such as `> VALUE1 and < VALUE2`.

For this example with program `COBIMSB` we will retrieve root segments for all employees whose `EMPID` is greater than 3000 and less than 7000. For that we simply need to create and use a new SSA. Here it is:

```
01 EMP-QUALIFIED-SSA-BOOL.
   05  SEGNAME     PIC X(08) VALUE 'EMPLOYEE'.
   05  FILLER      PIC X(01) VALUE '('.
   05  FIELD       PIC X(08) VALUE 'EMPID'.
   05  OPER        PIC X(02) VALUE '>='.
   05  EMP-ID-VAL1 PIC X(04) VALUE '    '.
   05  OPER        PIC X(01) VALUE '&'.
   05  FIELD2      PIC X(08) VALUE 'EMPID'.
   05  OPER2       PIC X(02) VALUE '<='.
   05  EMP-ID-VAL2 PIC X(04) VALUE '    '.
   05  FILLER      PIC X(01) VALUE ')'.
```

For the above we must load (or initialize) the minimum value 3000 into `EMP-ID-VAL1`,

131

and the ceiling value 7000 into `EMP-ID-VAL2`. Then we'll call the database using the `EMP-QUALIFIED-SSA-BOOL` SSA. We'll do a loop through the database and our retrieval loop should only return those employee records that satisfy the Boolean SSA.

Note that to end our read loop, we check both for IMS status codes `GB` and `GE`. This is because the last record that satisfies the database call may not be the physical end of the database. Consequently reading beyond the end of the "result set" of your database call will result in a `GE` status code unless it happens to also be the end of the database. So you have to check for both `GB` and `GE`.

Here is our program source code.

```
IDENTIFICATION DIVISION.
PROGRAM-ID. COBIMSB.

* * * * * * * * * * * * * * * * * * * * * * * * * * * * * * * * * * * * * * * * *
*   WALK THROUGH THE EMPLOYEE SEGMENTS OF THE ENTIRE *
*   EMPLOYEE IMS DATABASE USING BOOLEAN SSA.         *
* * * * * * * * * * * * * * * * * * * * * * * * * * * * * * * * * * * * * * * * *

ENVIRONMENT DIVISION.
INPUT-OUTPUT SECTION.
DATA DIVISION.

* * * * * * * * * * * * * * * * * * * * * * * * * * * * * * * * * * * * * * * * *
*  W O R K I N G   S T O R A G E   S E C T I O N    *
* * * * * * * * * * * * * * * * * * * * * * * * * * * * * * * * * * * * * * * * *

WORKING-STORAGE SECTION.

01 WS-FLAGS.
   05  SW-END-OF-DB-SWITCH     PIC X(1) VALUE 'N'.
       88  SW-END-OF-DB                 VALUE 'Y'.
       88  SW-NOT-END-OF-DB             VALUE 'N'.

01 DLI-FUNCTIONS.
   05 DLI-FUNCISRT  PIC X(4) VALUE 'ISRT'.
   05 DLI-FUNCGU    PIC X(4) VALUE 'GU  '.
   05 DLI-FUNCGN    PIC X(4) VALUE 'GN  '.
   05 DLI-FUNCGHU   PIC X(4) VALUE 'GHU '.
   05 DLI-FUNCGNP   PIC X(4) VALUE 'GNP '.
   05 DLI-FUNCREPL  PIC X(4) VALUE 'REPL'.
   05 DLI-FUNCDLET  PIC X(4) VALUE 'DLET'.
   05 DLI-FUNCXRST  PIC X(4) VALUE 'XRST'.
```

```
    05 DLI-FUNCCKPT   PIC X(4) VALUE 'CKPT'.

01 IO-EMPLOYEE-RECORD.
    05  EMPL-ID-IN    PIC X(04).
    05  FILLER        PIC X(01).
    05  EMPL-LNAME    PIC X(30).
    05  FILLER        PIC X(01).
    05  EMPL-FNAME    PIC X(20).
    05  FILLER        PIC X(01).
    05  EMPL-YRS-SRV  PIC X(02).
    05  FILLER        PIC X(01).
    05  EMPL-PRM-DTE  PIC X(10).
    05  FILLER        PIC X(10).

01 EMP-UNQUALIFIED-SSA.
    05  SEGNAME       PIC X(08) VALUE 'EMPLOYEE'.
    05  FILLER        PIC X(01) VALUE ' '.

01 EMP-QUALIFIED-SSA.
    05  SEGNAME       PIC X(08) VALUE 'EMPLOYEE'.
    05  FILLER        PIC X(01) VALUE '('.
    05  FIELD         PIC X(08) VALUE 'EMPID'.
    05  OPER          PIC X(02) VALUE ' ='.
    05  EMP-ID-VAL    PIC X(04) VALUE '    '.
    05  FILLER        PIC X(01) VALUE ')'.

01 EMP-QUALIFIED-SSA-BOOL.
    05  SEGNAME       PIC X(08) VALUE 'EMPLOYEE'.
    05  FILLER        PIC X(01) VALUE '('.
    05  FIELD         PIC X(08) VALUE 'EMPID'.
    05  OPER          PIC X(02) VALUE '>='.
    05  EMP-ID-VAL1   PIC X(04) VALUE '    '.
    05  OPER          PIC X(01) VALUE '&'.
    05  FIELD2        PIC X(08) VALUE 'EMPID'.
    05  OPER2         PIC X(02) VALUE '<='.
    05  EMP-ID-VAL2   PIC X(04) VALUE '    '.
    05  FILLER        PIC X(01) VALUE ')'.

01 SEG-IO-AREA      PIC X(80).

01 IMS-RET-CODES.
    05 THREE          PIC S9(9) COMP VALUE +3.
    05 FOUR           PIC S9(9) COMP VALUE +4.
    05 FIVE           PIC S9(9) COMP VALUE +5.
    05 SIX            PIC S9(9) COMP VALUE +6.


LINKAGE SECTION.
```

```
    01  PCB-MASK.
        03 DBD-NAME          PIC X(8).
        03 SEG-LEVEL         PIC XX.
        03 STATUS-CODE       PIC XX.
        03 PROC-OPT          PIC X(4).
        03 FILLER            PIC X(4).
        03 SEG-NAME          PIC X(8).
        03 KEY-FDBK          PIC S9(5) COMP.
        03 NUM-SENSEG        PIC S9(5) COMP.
        03 KEY-FDBK-AREA.
           05 EMPLOYEE-KEY  PIC X(04).
           05 EMPPAYHS-KEY  PIC X(08).

  PROCEDURE DIVISION.

      INITIALIZE PCB-MASK
      ENTRY 'DLITCBL' USING PCB-MASK

      PERFORM P100-INITIALIZATION.
      PERFORM P200-MAINLINE.
      PERFORM P300-TERMINATION.
      GOBACK.

  P100-INITIALIZATION.

      DISPLAY '** PROGRAM COBIMSB START **'
      DISPLAY 'PROCESSING IN P100-INITIALIZATION'.
      MOVE '3000' TO EMP-ID-VAL1
      MOVE '7000' TO EMP-ID-VAL2

*     DO INITIAL DB READ FOR FIRST EMPLOYEE RECORD

      CALL 'CBLTDLI' USING FOUR,
           DLI-FUNCGN,
           PCB-MASK,
           SEG-IO-AREA,
           EMP-QUALIFIED-SSA-BOOL

      IF STATUS-CODE = '  ' THEN
         NEXT SENTENCE
      ELSE
         IF STATUS-CODE = 'GE' OR
            STATUS-CODE = 'GB' THEN
            SET SW-END-OF-DB TO TRUE
            DISPLAY 'END OF DATABASE :'
         ELSE
            PERFORM P400-DISPLAY-ERROR
            GOBACK
```

```
            END-IF

        END-IF.

    P200-MAINLINE.

        DISPLAY 'PROCESSING IN P200-MAINLINE'

*       CHECK STATUS CODE AND FIRST RECORD

        IF SW-END-OF-DB THEN
            DISPLAY 'NO RECORDS TO PROCESS!!'
        ELSE
            PERFORM UNTIL SW-END-OF-DB
                DISPLAY 'SUCCESSFUL READ :' SEG-IO-AREA
                CALL 'CBLTDLI' USING FOUR,
                    DLI-FUNCGN,
                    PCB-MASK,
                    SEG-IO-AREA,
                    EMP-QUALIFIED-SSA-BOOL

                IF STATUS-CODE = 'GB' OR 'GE' THEN
                    SET SW-END-OF-DB TO TRUE
                    DISPLAY 'END OF DATABASE'
                ELSE
                    IF STATUS-CODE NOT EQUAL SPACES THEN
                        PERFORM P400-DISPLAY-ERROR
                        GOBACK
                    END-IF
                END-IF

            END-PERFORM.

        DISPLAY 'FINISHED PROCESSING IN P200-MAINLINE'.

    P300-TERMINATION.

        DISPLAY 'PROCESSING IN P300-TERMINATION'
        DISPLAY '** COBIMSB - SUCCESSFULLY ENDED **'.

    P400-DISPLAY-ERROR.

        DISPLAY 'ERROR ENCOUNTERED - DETAIL FOLLOWS'
        DISPLAY 'SEG-IO-AREA    :' SEG-IO-AREA
        DISPLAY 'DBD-NAME1:'      DBD-NAME
        DISPLAY 'SEG-LEVEL1:'     SEG-LEVEL
        DISPLAY 'STATUS-CODE:'    STATUS-CODE
        DISPLAY 'PROC-OPT1 :'     PROC-OPT
```

135

```
            DISPLAY 'SEG-NAME1 :'      SEG-NAME
            DISPLAY 'KEY-FDBK1 :'      KEY-FDBK
            DISPLAY 'NUM-SENSEG1:'     NUM-SENSEG
            DISPLAY 'KEY-FDBK-AREA1:' KEY-FDBK-AREA.

      *     END OF SOURCE CODE
```

After we compile, link and execute, here is the output. As you can see, the only employees retrieved are those whose ids fall between 3,000 and 7,000 inclusive.

```
** PROGRAM COBIMSB START **
PROCESSING IN P100-INITIALIZATION
EMP-QUALIFIED-SSA-BOOL EMPLOYEE(EMPID   >=3000&EMPID   <=7000)
PROCESSING IN P200-MAINLINE
SUCCESSFUL READ :3217 JOHNSON                EDWARD              04 201
SUCCESSFUL READ :4175 TURNBULL               FRED                01 201
SUCCESSFUL READ :4720 SCHULTZ                TIM                 09 201
SUCCESSFUL READ :4836 SMITH                  SANDRA              03 201
SUCCESSFUL READ :6288 WILLARD                JOE                 06 201
END OF DATABASE
FINISHED PROCESSING IN P200-MAINLINE
PROCESSING IN P300-TERMINATION
** COBIMSB - SUCCESSFULLY ENDED **
```

Extended Boolean SSAs can be very handy when you need to ready a range of values, or for any retrieval that must satisfy multiple conditions.

Here's the PLI source code for this example.

```
PLIIMSB: PROCEDURE (DB_PTR_PCB) OPTIONS(MAIN);
/****************************************************************
* PROGRAM NAME :   PLIIMSB - WALK THROUGH THE ROOT SEGMENTS OF   *
*                  EMPLOYEE DB USING BOOLEAN SSA.                *
****************************************************************/
/****************************************************************
/*              W O R K I N G   S T O R A G E                   *
****************************************************************/

   DCL ONCODE                  BUILTIN;
   DCL DB_PTR_PCB              POINTER;
   DCL PLITDLI                 EXTERNAL ENTRY;
   DCL SW_END_OF_DB            STATIC BIT(01) INIT('0'B);

   DCL 01 DLI_FUNCTIONS,
        05 DLI_FUNCISRT        CHAR(04) INIT ('ISRT'),
        05 DLI_FUNCGU          CHAR(04) INIT ('GU  '),
        05 DLI_FUNCGN          CHAR(04) INIT ('GN  '),
        05 DLI_FUNCGHU         CHAR(04) INIT ('GHU '),
        05 DLI_FUNCGNP         CHAR(04) INIT ('GNP '),
```

```
            05 DLI_FUNCREPL          CHAR(04) INIT ('REPL'),
            05 DLI_FUNCDLET          CHAR(04) INIT ('DLET'),
            05 DLI_FUNCXRST          CHAR(04) INIT ('XRST'),
            05 DLI_FUNCCHKP          CHAR(04) INIT ('CHKP'),
            05 DLI_FUNCROLL          CHAR(04) INIT ('ROLL');


    DCL 01 IO_EMPLOYEE_RECORD,
            05  EMPL_ID_IN           CHAR(04),
            05  FILLER1              CHAR(01),
            05  EMPL_LNAME           CHAR(30),
            05  FILLER2              CHAR(01),
            05  EMPL_FNAME           CHAR(20),
            05  FILLER3              CHAR(01),
            05  EMPL_YRS_SRV         CHAR(02),
            05  FILLER4              CHAR(01),
            05  EMPL_PRM_DTE         CHAR(10),
            05  FILLER5              CHAR(01),
            05  EMPL_SSN             CHAR(09);

    DCL 01 PCB_MASK                  BASED(DB_PTR_PCB),
            05 DBD_NAME              CHAR(08),
            05 SEG_LEVEL             CHAR(02),
            05 STATUS_CODE           CHAR(02),
            05 PROC_OPT              CHAR(04),
            05 FILLER6               FIXED BIN (31),
            05 SEG_NAME              CHAR(08),
            05 KEY_FDBK              FIXED BIN (31),
            05 NUM_SENSEG            FIXED BIN (31),
            05 KEY_FDBK_AREA,
               10 EMPLOYEE_ID        CHAR(04);

    DCL 01 EMP_UNQUALIFIED_SSA,
            05  SEGNAME              CHAR(08) INIT ('EMPLOYEE'),
            05  FILLER7              CHAR(01) INIT (' ');

    DCL 01 EMP_QUALIFIED_SSA,
            05  SEGNAME              CHAR(08) INIT('EMPLOYEE'),
            05  FILLER8              CHAR(01) INIT('('),
            05  FIELD                CHAR(08) INIT('EMPID'),
            05  OPER                 CHAR(02) INIT(' ='),
            05  EMP_ID_VAL           CHAR(04) INIT('    '),
            05  FILLER9              CHAR(01) INIT(')');

    DCL 01 EMP_QUALIFIED_SSA_BOOL,
            05  SEGNAME              CHAR(08) INIT ('EMPLOYEE'),
            05  FILLER10             CHAR(01) INIT('('),
            05  FIELD                CHAR(08) INIT('EMPID'),
            05  OPER                 CHAR(02) INIT('>='),
            05  EMP_ID_VAL1          CHAR(04) INIT('    '),
            05  OPER2                CHAR(01) INIT('&'),
            05  FIELD2               CHAR(08) INIT('EMPID'),
            05  OPER3                CHAR(02) INIT('<='),
            05  EMP_ID_VAL2          CHAR(04) INIT('    '),
            05  FILLER11             CHAR(01) INIT(')');
```

```
    DCL THREE                    FIXED BIN (31) INIT(3);
    DCL FOUR                     FIXED BIN (31) INIT(4);
    DCL FIVE                     FIXED BIN (31) INIT(5);
    DCL SIX                      FIXED BIN (31) INIT(6);

/*******************************************************************
/*                P R O G R A M   M A I N L I N E            *
*******************************************************************/
CALL P100_INITIALIZATION;
CALL P200_MAINLINE;
CALL P300_TERMINATION;

P100_INITIALIZATION: PROC;

    PUT SKIP LIST ('PLIIMSB: TRAVERSE EMPLOYEE DB ROOT SEGS USING BOOL
        SSAs');
    PCB_MASK = '';
    IO_EMPLOYEE_RECORD  = '';

  /*  SET THE EMPLOYEE SEGMENT SEARCH ARGUMENT */

    EMP_ID_VAL1 = '3000';
    EMP_ID_VAL2 = '7000';

 /* DO INITIAL DB READ FOR FIRST EMPLOYEE RECORD */

    CALL PLITDLI (FOUR,
                  DLI_FUNCGN,
                  PCB_MASK,
                  IO_EMPLOYEE_RECORD,
                  EMP_QUALIFIED_SSA_BOOL);

    IF STATUS_CODE = '  ' THEN;
    ELSE
       IF STATUS_CODE = 'GB' |
          STATUS_CODE = 'GE' THEN
          DO;
             SW_END_OF_DB = '1'B;
             PUT SKIP LIST ('** END OF DATABASE');
          END;
       ELSE
          DO;
             CALL P400_DISPLAY_ERROR;
             RETURN;
          END;

END P100_INITIALIZATION;

P200_MAINLINE: PROC;

    /*  MAIN LOOP - CYCLE THROUGH ALL ROOT SEGMENTS IN THE DB,
                    DISPLAYING THE DATA RETRIEVED                */
```

```
        IF SW_END_OF_DB THEN PUT SKIP LIST ('NO RECORDS TO PROCESS!!');
        ELSE
            DO WHILE (¬SW_END_OF_DB);
                PUT SKIP LIST ('SUCCESSFUL READ USING BOOLEAN SSA : ');
                PUT SKIP DATA (IO_EMPLOYEE_RECORD);

                CALL PLITDLI (FOUR,
                              DLI_FUNCGN,
                              PCB_MASK,
                              IO_EMPLOYEE_RECORD,
                              EMP_QUALIFIED_SSA_BOOL);

                IF STATUS_CODE = '  ' THEN;
                ELSE
                    IF STATUS_CODE = 'GB' | STATUS_CODE = 'GE' THEN
                        DO;
                            SW_END_OF_DB = '1'B;
                            PUT SKIP LIST ('** END OF DATABASE');
                        END;
                    ELSE
                        DO;
                            CALL P400_DISPLAY_ERROR;
                            RETURN;
                        END;

            END; /* DO WHILE */

        PUT SKIP LIST ('FINISHED PROCESSING IN P200_MAINLINE');

END P200_MAINLINE;

P300_TERMINATION: PROC;

        PUT SKIP LIST ('PLIIMSB - ENDED SUCCESSFULLY');

END P300_TERMINATION;

P400_DISPLAY_ERROR: PROC;

        PUT SKIP LIST ('ERROR ENCOUNTERED - DETAIL FOLLOWS');
        PUT SKIP LIST ('IO_EMPLOYEE_RECORD :'
            || IO_EMPLOYEE_RECORD);
        PUT SKIP LIST ('DBD_NAME1:' || DBD_NAME);
        PUT SKIP LIST ('SEG_LEVEL1:' || SEG_LEVEL);
        PUT SKIP LIST ('STATUS_CODE:' || STATUS_CODE);
        PUT SKIP LIST ('PROC_OPT1 :' || PROC_OPT);
        PUT SKIP LIST ('SEG_NAME1 :' || SEG_NAME);
        PUT SKIP LIST ('KEY_FDBK1 :' || KEY_FDBK);
        PUT SKIP LIST ('NUM_SENSEG1:' || NUM_SENSEG);
        PUT SKIP LIST ('KEY_FDBK_AREA1:' || KEY_FDBK_AREA);

END P400_DISPLAY_ERROR;

END PLIIMSB;
```

Command Codes

IMS command codes change and/or extend the way an IMS call works. There are about 18 command codes that serve various purposes. See the table at the end of this topic for all the command codes and what they do.

We'll do an example of the C command code. The C command code allows you to issue a qualified SSA using the concatenated key for a child segment rather than using separate SSAs for the various parent/child segments. For example suppose we want to retrieve the paycheck record of employee 3217 for pay effective January 1, 2017, and for payday February 15, 2017. The concatenated key for that is as follows:

```
321720170101201702015
```

This is the key for the root segment (3217) plus the key for the EMPPAY segment (20170101), plus the key for the EMPPAYHS segment (20170215).

To use the C command code, we must create a new SSA structure that uses both the C command code, and accommodates the concatenated key. It will look like this:

```
01 EMPPAYHS-CCODE-SSA.
   05  SEGNAME     PIC X(08) VALUE 'EMPPAYHS'.
   05  FILLER      PIC X(02) VALUE '*C'.
   05  FILLER      PIC X(01) VALUE '('.
   05  CONCATKEY   PIC X(20) VALUE SPACES.
   05  FILLER      PIC X(01) VALUE ')'.
```

Like all SSAs, our new one includes the segment name. Position 9 of the SSA will contain an asterisk (*) or blank if a command code is not being used. We put a C in position 10 to indicate we are using a concatenated key command code. We've named our concatenated key variable CONCATKEY (the name is arbitrary – you could use any name for this variable).

The CONCATKEY length is 20 bytes (4 for the employee id, and 8 each for the salary effective date and the pay date. We have initialized the concatenated key variable to the value we are looking for. You could also load it using a MOVE statement.

Ok here is the complete code for COBIMSC. It should look very familiar except for the SSA. For comparison, we will first use the regular multiple SSA method to call the

2/15 pay record. Then we will use a second call with the C command code method and a concatenated key. The results should be identical.

```
IDENTIFICATION DIVISION.
PROGRAM-ID. COBIMSC.

*********************************************************
*    READ AND DISPLAY EMP HISTORY RECORD FROM          *
*    EMPLOYEE IMS DATABASE. THIS EXAMPLE USES A         *
*    C COMMAND CODE TO PROVIDE THE CONCATENATED         *
*    KEY SSA RATHER THAN A QUALIFICATION STATEMENT      *
*    SSA (second example).                              *
*********************************************************

ENVIRONMENT DIVISION.
DATA DIVISION.

*********************************************************
*  W O R K I N G   S T O R A G E   S E C T I O N   *
*********************************************************

WORKING-STORAGE SECTION.

01 WS-FLAGS.
    05  SW-END-OF-DB-SWITCH      PIC X(1) VALUE 'N'.
        88  SW-END-OF-DB                  VALUE 'Y'.
        88  SW-NOT-END-OF-DB              VALUE 'N'.
    05  SW-END-OF-EMPPAYHS-SW    PIC X(1) VALUE 'N'.
        88  SW-END-OF-EMPPAYHS            VALUE 'Y'.
        88  SW-NOT-END-OF-EMPPAYHS        VALUE 'N'.

01 IO-EMPLOYEE-RECORD.
    05  EMPL-ID        PIC X(04).
    05  FILLER         PIC X(01).
    05  EMPL-LNAME     PIC X(30).
    05  FILLER         PIC X(01).
    05  EMPL-FNAME     PIC X(20).
    05  FILLER         PIC X(01).
    05  EMPL-YRS-SRV   PIC X(02).
    05  FILLER         PIC X(01).
    05  EMPL-PRM-DTE   PIC X(10).
    05  FILLER         PIC X(10).

01 IO-EMPPAY-RECORD.
    05  PAY-EFF-DATE   PIC X(8).
    05  PAY-REG-PAY    PIC S9(6)V9(2) USAGE COMP-3.
    05  PAY-BON-PAY    PIC S9(6)V9(2) USAGE COMP-3.
    05  SEMIMTH-PAY    PIC S9(6)V9(2) USAGE COMP-3.
    05  FILLER         PIC X(57).
```

141

```
01 IO-EMPPAYHS-RECORD.
   05   PAY-DATE       PIC X(8).
   05   PAY-ANN-PAY    PIC S9(6)V9(2) USAGE COMP-3.
   05   PAY-AMT        PIC S9(6)V9(2) USAGE COMP-3.
   05   FILLER         PIC X(62).

01 SEG-IO-AREA         PIC X(80).

01 IMS-RET-CODES.
   05 THREE            PIC S9(9) COMP VALUE +3.
   05 FOUR             PIC S9(9) COMP VALUE +4.
   05 FIVE             PIC S9(9) COMP VALUE +5.
   05 SIX              PIC S9(9) COMP VALUE +6.

01 DLI-FUNCTIONS.
   05 DLI-FUNCISRT     PIC X(4) VALUE 'ISRT'.
   05 DLI-FUNCGU       PIC X(4) VALUE 'GU  '.
   05 DLI-FUNCGN       PIC X(4) VALUE 'GN  '.
   05 DLI-FUNCGHU      PIC X(4) VALUE 'GHU '.
   05 DLI-FUNCGNP      PIC X(4) VALUE 'GNP '.
   05 DLI-FUNCREPL     PIC X(4) VALUE 'REPL'.
   05 DLI-FUNCDLET     PIC X(4) VALUE 'DLET'.
   05 DLI-FUNCXRST     PIC X(4) VALUE 'XRST'.
   05 DLI-FUNCCKPT     PIC X(4) VALUE 'CKPT'.

01 DISPLAY-EMPPAYHS-PIC.
   05  DIS-REG-PAY    PIC ZZ999.99-.
   05  DIS-SMT-PAY    PIC ZZ999.99-.

 01 EMP-UNQUALIFIED-SSA.
   05  SEGNAME        PIC X(08) VALUE 'EMPLOYEE'.
   05  FILLER         PIC X(01) VALUE ' '.

 01 EMP-QUALIFIED-SSA.
   05  SEGNAME        PIC X(08) VALUE 'EMPLOYEE'.
   05  FILLER         PIC X(01) VALUE '('.
   05  FIELD          PIC X(08) VALUE 'EMPID'.
   05  OPER           PIC X(02) VALUE ' ='.
   05  EMP-ID-VAL     PIC X(04) VALUE '    '.
   05  FILLER         PIC X(01) VALUE ')'.

 01 EMPPAY-UNQUALIFIED-SSA.
   05  SEGNAME        PIC X(08) VALUE 'EMPPAY  '.
   05  FILLER         PIC X(01) VALUE ' '.

 01 EMPPAY-QUALIFIED-SSA.
   05  SEGNAME        PIC X(08) VALUE 'EMPPAY  '.
   05  FILLER         PIC X(01) VALUE '('.
   05  FIELD          PIC X(08) VALUE 'EFFDATE '.
```

```
          05   OPER         PIC X(02) VALUE ' ='.
          05   EFFDATE-VAL  PIC X(08) VALUE '          '.
          05   FILLER       PIC X(01) VALUE ')'.

     01 EMPPAYHS-UNQUALIFIED-SSA.
          05   SEGNAME      PIC X(08) VALUE 'EMPPAYHS'.
          05   FILLER       PIC X(01) VALUE ' '.

     01 EMPPAYHS-QUALIFIED-SSA.
          05   SEGNAME      PIC X(08) VALUE 'EMPPAYHS'.
          05   FILLER       PIC X(01) VALUE '('.
          05   FIELD        PIC X(08) VALUE 'PAYDATE '.
          05   OPER         PIC X(02) VALUE ' ='.
          05   PAYDATE-VAL  PIC X(08) VALUE '          '.
          05   FILLER       PIC X(01) VALUE ')'.

     01 EMPPAYHS-CCODE-SSA.
          05   SEGNAME      PIC X(08) VALUE 'EMPPAYHS'.
          05   FILLER       PIC X(02) VALUE '*C'.
          05   FILLER       PIC X(01) VALUE '('.
          05   CONCATKEY    PIC X(20) VALUE '32172017010120170215'.
          05   FILLER       PIC X(01) VALUE ')'.


 LINKAGE SECTION.
  01 PCB-MASK.
          03 DBD-NAME       PIC X(8).
          03 SEG-LEVEL      PIC XX.
          03 STATUS-CODE    PIC XX.
          03 PROC-OPT       PIC X(4).
          03 FILLER         PIC X(4).
          03 SEG-NAME       PIC X(8).
          03 KEY-FDBK       PIC S9(5) COMP.
          03 NUM-SENSEG     PIC S9(5) COMP.
          03 KEY-FDBK-AREA.
             05 EMPLOYEE-ID  PIC X(04).
             05 EMPPAYHS     PIC X(08).

 PROCEDURE DIVISION.

     INITIALIZE PCB-MASK
     ENTRY 'DLITCBL' USING PCB-MASK

     PERFORM P100-INITIALIZATION.
     PERFORM P200-MAINLINE.
     PERFORM P300-TERMINATION.
     GOBACK.
```

```
P100-INITIALIZATION.

    DISPLAY '** PROGRAM COBIMSC START **'
    DISPLAY 'PROCESSING IN P100-INITIALIZATION'.


P200-MAINLINE.

    DISPLAY 'PROCESSING IN P200-MAINLINE'

    MOVE '3217'     TO EMP-ID-VAL
    MOVE '20170101' TO EFFDATE-VAL
    MOVE '201700215' TO PAYDATE-VAL

    CALL 'CBLTDLI' USING SIX,
         DLI-FUNCGU,
         PCB-MASK,
         IO-EMPPAYHS-RECORD,
         EMP-QUALIFIED-SSA,
         EMPPAY-QUALIFIED-SSA,
         EMPPAYHS-QUALIFIED-SSA.

    EVALUATE STATUS-CODE
       WHEN ' '
          DISPLAY 'GOOD READ OF EMPPAYHS : '
              EMP-ID-VAL
          MOVE PAY-ANN-PAY TO DIS-REG-PAY
          MOVE PAY-AMT     TO DIS-SMT-PAY
          DISPLAY 'PAY-DATE   : ' PAY-DATE
          DISPLAY 'PAY-ANN-PAY: ' DIS-REG-PAY
          DISPLAY 'PAY-AMT    : ' DIS-SMT-PAY
       WHEN 'GE'
       WHEN 'GB'
          DISPLAY 'PAY HISTORY SEGMENT NOT FOUND'
       WHEN OTHER
          PERFORM P9000-DISPLAY-ERROR
          GOBACK
     END-EVALUATE.

    DISPLAY 'NOW CALLING THE 2/15/2017 REC USING C COMMAND CODE'

    CALL 'CBLTDLI' USING FOUR,
         DLI-FUNCGU,
         PCB-MASK,
         IO-EMPPAYHS-RECORD,
         EMPPAYHS-CCODE-SSA.

    EVALUATE STATUS-CODE
       WHEN ' '
          DISPLAY 'GOOD READ OF EMPPAYHS : '
```

```
            EMP-ID-VAL
        MOVE PAY-ANN-PAY TO DIS-REG-PAY
        MOVE PAY-AMT     TO DIS-SMT-PAY
        DISPLAY 'PAY-DATE    : ' PAY-DATE
        DISPLAY 'PAY-ANN-PAY: ' DIS-REG-PAY
        DISPLAY 'PAY-AMT     : ' DIS-SMT-PAY
    WHEN 'GE'
    WHEN 'GB'
        DISPLAY 'PAY HISTORY SEGMENT NOT FOUND'
    WHEN OTHER
        PERFORM P9000-DISPLAY-ERROR
        GOBACK
    END-EVALUATE.

    DISPLAY 'FINISHED PROCESSING IN P200-MAINLINE'.

P300-TERMINATION.

    DISPLAY 'PROCESSING IN P300-TERMINATION'
    DISPLAY '** COBIMSC - SUCCESSFULLY ENDED **'.

P9000-DISPLAY-ERROR.

    DISPLAY 'ERROR ENCOUNTERED - DETAIL FOLLOWS'
    DISPLAY 'DBD-NAME1:'     DBD-NAME
    DISPLAY 'SEG-LEVEL1:'    SEG-LEVEL
    DISPLAY 'STATUS-CODE:'   STATUS-CODE
    DISPLAY 'PROC-OPT1 :'    PROC-OPT
    DISPLAY 'SEG-NAME1 :'    SEG-NAME
    DISPLAY 'KEY-FDBK1 :'    KEY-FDBK
    DISPLAY 'NUM-SENSEG1:'   NUM-SENSEG
    DISPLAY 'KEY-FDBK-AREA1:' KEY-FDBK-AREA.

*   END OF SOURCE CODE
```

Ok, once again we compile, link and execute. Here is our output.

```
** PROGRAM COBIMSC START **
PROCESSING IN P100-INITIALIZATION
PROCESSING IN P200-MAINLINE
FIRST CALL THE 2/15/2017 PAY REC WITH 3 SSA METHOD
GOOD READ OF EMPPAYHS : 3217
PAY-DATE    : 20170215
PAY-ANN-PAY: 65000.00
PAY-AMT     : 2708.33
NOW CALLING THE 2/15/2017 REC USING C COMMAND CODE
GOOD READ OF EMPPAYHS : 3217
PAY-DATE    : 20170215
PAY-ANN-PAY: 65000.00
```

145

```
PAY-AMT    :  2708.33
FINISHED PROCESSING IN P200-MAINLINE
PROCESSING IN P300-TERMINATION
** COBIMSC - SUCCESSFULLY ENDED **
```

PLI source code for this example is here:

```
PLIIMSC: PROCEDURE (DB_PTR_PCB) OPTIONS(MAIN);
/********************************************************************
* PROGRAM NAME :   PLIIMSC - RETRIEVE A RECORD FROM EMPLOYEE DB   *
*                            USING C COMMAND CODE TO PROVIDE A    *
*                            CONCATENATED KEY SSA.                *
********************************************************************/


/********************************************************************
/*               W O R K I N G   S T O R A G E                    *
********************************************************************/

    DCL ONCODE                  BUILTIN;
    DCL DB_PTR_PCB              POINTER;
    DCL PLITDLI                 EXTERNAL ENTRY;

    DCL 01 DLI_FUNCTIONS,
           05 DLI_FUNCISRT       CHAR(04) INIT ('ISRT'),
           05 DLI_FUNCGU         CHAR(04) INIT ('GU  '),
           05 DLI_FUNCGN         CHAR(04) INIT ('GN  '),
           05 DLI_FUNCGHU        CHAR(04) INIT ('GHU '),
           05 DLI_FUNCGNP        CHAR(04) INIT ('GNP '),
           05 DLI_FUNCREPL       CHAR(04) INIT ('REPL'),
           05 DLI_FUNCDLET       CHAR(04) INIT ('DLET'),
           05 DLI_FUNCXRST       CHAR(04) INIT ('XRST'),
           05 DLI_FUNCCHKP       CHAR(04) INIT ('CHKP'),
           05 DLI_FUNCROLL       CHAR(04) INIT ('ROLL');

    DCL 01 IO_EMPLOYEE_RECORD,
           05  EMPL_ID_IN        CHAR(04),
           05  FILLER1           CHAR(01),
           05  EMPL_LNAME        CHAR(30),
           05  FILLER2           CHAR(01),
           05  EMPL_FNAME        CHAR(20),
           05  FILLER3           CHAR(01),
           05  EMPL_YRS_SRV      CHAR(02),
           05  FILLER4           CHAR(01),
           05  EMPL_PRM_DTE      CHAR(10),
           05  FILLER5           CHAR(01),
           05  EMPL_SSN          CHAR(09);
```

```
DCL 01 IO_EMPPAY_RECORD,
       05  PAY_EFF_DATE       CHAR(8),
       05  PAY_REG_PAY        FIXED DEC (8,2),
       05  PAY_BON_PAY        FIXED DEC (8,2),
       05  SEMIMTH_PAY        FIXED DEC (8,2),
       05  FILLER6            CHAR(57);

DCL 01 IO_EMPPAYHS_RECORD,
       05  PAY_DATE           CHAR(8),
       05  PAY_ANN_PAY        FIXED DEC (8,2),
       05  PAY_AMT            FIXED DEC (8,2),
       05  FILLER65           CHAR(62);

DCL 01 PCB_MASK               BASED(DB_PTR_PCB),
       05 DBD_NAME            CHAR(08),
       05 SEG_LEVEL           CHAR(02),
       05 STATUS_CODE         CHAR(02),
       05 PROC_OPT            CHAR(04),
       05 FILLER99            FIXED BIN (31),
       05 SEG_NAME            CHAR(08),
       05 KEY_FDBK            FIXED BIN (31),
       05 NUM_SENSEG          FIXED BIN (31),
       05 KEY_FDBK_AREA,
          10 EMPLOYEE_ID      CHAR(04);

DCL 01 EMP_UNQUALIFIED_SSA,
       05  SEGNAME            CHAR(08) INIT ('EMPLOYEE'),
       05  FILLER7            CHAR(01) INIT (' ');

DCL 01 EMP_QUALIFIED_SSA,
       05  SEGNAME            CHAR(08) INIT('EMPLOYEE'),
       05  FILLER8            CHAR(01) INIT('('),
       05  FIELD              CHAR(08) INIT('EMPID'),
       05  OPER               CHAR(02) INIT(' ='),
       05  EMP_ID_VAL         CHAR(04) INIT('    '),
       05  FILLER9            CHAR(01) INIT(')');

DCL 01 EMPPAY_UNQUALIFIED_SSA,
       05  SEGNAME            CHAR(08) INIT('EMPPAY  '),
       05  FILLER10           CHAR(01) INIT(' ');

DCL 01 EMPPAY_QUALIFIED_SSA,
       05  SEGNAME            CHAR(08) INIT('EMPPAY  '),
       05  FILLER11           CHAR(01) INIT('('),
       05  FIELD              CHAR(08) INIT('EFFDATE '),
       05  OPER               CHAR(02) INIT(' ='),
       05  EFFDATE_VAL        CHAR(08) INIT('        '),
       05  FILLER12           CHAR(01) INIT(')');
```

```
     DCL 01 EMP_QUALIFIED_SSA_EMPSSN,
          05  SEGNAME          CHAR(08) INIT('EMPLOYEE'),
          05  FILLER10         CHAR(01) INIT('('),
          05  FIELD            CHAR(08) INIT('EMPSSN'),
          05  OPER             CHAR(02) INIT(' ='),
          05  EMPSSN_VAL       CHAR(09) INIT('         '),
          05  FILLER11         CHAR(01) INIT(')');

     DCL 01 EMPPAYHS_UNQUALIFIED_SSA,
          05  SEGNAME          CHAR(08) INIT('EMPPAYHS'),
          05  FILLER12         CHAR(01) INIT(' ');

     DCL 01 EMPPAYHS_QUALIFIED_SSA,
          05  SEGNAME          CHAR(08) INIT('EMPPAYHS'),
          05  FILLER13         CHAR(01) INIT('('),
          05  FIELD            CHAR(08) INIT('PAYDATE '),
          05  OPER             CHAR(02) INIT(' ='),
          05  PAYDATE_VAL      CHAR(08) INIT('        '),
          05  FILLER14         CHAR(01) INIT(')');

     DCL 01 EMPPAYHS_CCODE_SSA,
          05  SEGNAME          CHAR(08) INIT('EMPPAYHS'),
          05  FILLER15         CHAR(02) INIT('*C'),
          05  FILLER16         CHAR(01) INIT('('),
          05  CONCATKEY        CHAR(20) INIT('                    '),
          05  FILLER17         CHAR(01) INIT(')');

  DCL THREE                    FIXED BIN (31) INIT(3);
  DCL FOUR                     FIXED BIN (31) INIT(4);
  DCL FIVE                     FIXED BIN (31) INIT(5);
  DCL SIX                      FIXED BIN (31) INIT(6);
/****************************************************************
/*            P R O G R A M   M A I N L I N E                  *
****************************************************************/

CALL P100_INITIALIZATION;
CALL P200_MAINLINE;
CALL P300_TERMINATION;

P100_INITIALIZATION: PROC;

   PUT SKIP LIST ('PLIIMSC: GET EMPPAYHS REC FROM DB USING CMD CODE');
   PUT SKIP LIST ('PROCESSING IN P100_INITIALIZATION');
   IO_EMPLOYEE_RECORD  = '';
   IO_EMPPAY_RECORD    = '';
   IO_EMPPAYHS_RECORD  = '';
```

```
END P100_INITIALIZATION;

P200_MAINLINE: PROC;

    /*  SET THE EMPLOYEE SEGMENT SEARCH ARGUMENT AND CALL PLITDLI */

    EMP_ID_VAL  = '3217';
    EFFDATE_VAL = '20170101';
    PAYDATE_VAL = '20170215';

    PUT SKIP LIST ('1ST CALL THE 2/15/2017 PAY REC WITH 3 SSA METHOD');

    CALL PLITDLI (SIX,
                  DLI_FUNCGU,
                  PCB_MASK,
                  IO_EMPPAYHS_RECORD,
                  EMP_QUALIFIED_SSA,
                  EMPPAY_QUALIFIED_SSA,
                  EMPPAYHS_QUALIFIED_SSA);

    SELECT (STATUS_CODE);
       WHEN ('  ')
          DO;
             PUT SKIP LIST ('GOOD READ OF EMPPAYHS : ' || EMP_ID_VAL);
             PUT SKIP DATA (IO_EMPPAYHS_RECORD);
          END;
       WHEN ('GE','GB')
          PUT SKIP LIST ('PAY HIST SEG NOT FOUND FOR ' || EMP_ID_VAL);
       OTHERWISE
          DO;
             CALL P400_DISPLAY_ERROR;
             RETURN;
          END;

     END; /* SELECT */

    PUT SKIP LIST ('2ND CALL THE 2/15/2017 PAY REC WITH CMD CODE C');

    CONCATKEY = '32172017010120170215';

    CALL PLITDLI (FOUR,
                  DLI_FUNCGU,
                  PCB_MASK,
                  IO_EMPPAYHS_RECORD,
                  EMPPAYHS_CCODE_SSA);

    SELECT (STATUS_CODE);
       WHEN ('  ')
```

149

```
                DO;
                    PUT SKIP LIST ('GOOD READ OF EMPPAYHS : ' || EMP_ID_VAL);
                    PUT SKIP DATA (IO_EMPPAYHS_RECORD);
                END;
            WHEN ('GE','GB')
                PUT SKIP LIST ('PAY HIST SEG NOT FOUND FOR ' || EMP_ID_VAL);
            OTHERWISE
                DO;
                    CALL P400_DISPLAY_ERROR;
                    RETURN;
                END;

        END; /* SELECT */

END P200_MAINLINE;

P300_TERMINATION: PROC;

    PUT SKIP LIST ('PLIIMSC - ENDED SUCCESSFULLY');

END P300_TERMINATION;

P400_DISPLAY_ERROR: PROC;

    PUT SKIP LIST ('ERROR ENCOUNTERED - DETAIL FOLLOWS');
    PUT SKIP LIST ('IO_EMPPAYHS_RECORD: ' || IO_EMPLOYEE_RECORD);
    PUT SKIP LIST ('DBD_NAME1:' || DBD_NAME);
    PUT SKIP LIST ('SEG_LEVEL1:' || SEG_LEVEL);
    PUT SKIP LIST ('STATUS_CODE:' || STATUS_CODE);
    PUT SKIP LIST ('PROC_OPT1 :' || PROC_OPT);
    PUT SKIP LIST ('SEG_NAME1 :' || SEG_NAME);
    PUT SKIP LIST ('KEY_FDBK1 :' || KEY_FDBK);
    PUT SKIP LIST ('NUM_SENSEG1:' || NUM_SENSEG);
    PUT SKIP LIST ('KEY_FDBK_AREA1:' || KEY_FDBK_AREA);

END P400_DISPLAY_ERROR;

END PLIIMSC;
```

Command codes can be very useful when you need the features they offer. Check out the following table of the command codes and how they are used. This information is from the IBM product web site. [4] You'll find more detail about each command code there as well.

---

[4] https://www.ibm.com/support/knowledgecenter/en/SSEPH2_13.1.0/com.ibm.ims13.doc.apr/ims_cmd-codref.htm

Summary of Command Codes

Command Code	Description
A	Clear positioning and start the call at the beginning of the database.
C	Use the concatenated key of a segment to identify the segment.
D	Retrieve or insert a sequence of segments in a hierarchic path using only one call, instead of using a separate (path) call for each segment.
F	Back up to the first occurrence of a segment under its parent when searching for a particular segment occurrence. Disregarded for a root segment.
G	Prevent randomization or the calling of the HALDB Partition Selection exit routine and search the database sequentially.
L	Retrieve the last occurrence of a segment under its parent.
M	Move a subset pointer to the next segment occurrence after your current position. (Used with DEDBs only.)
N	Designate segments that you do not want replaced when replacing segments after a Get Hold call. Typically used when replacing a path of segments.
O	Either field names or both segment position and lengths can be contained in the SSA qualification for combine field position.
P	Set parentage at a higher level than what it usually is (the lowest-level SSA of the call).
Q	Reserve a segment so that other programs cannot update it until you have finished processing and updating it.
R	Retrieve the first segment occurrence in a subset. (Used with DEDBs only.)
S	Unconditionally set a subset pointer to the current position. (Used with DEDBs only.)

Command Code	Description
U	Limit the search for a segment to the dependents of the segment occurrence on which position is established.
V	Use the hierarchic level at the current position and higher as qualification for the segment.
W	Set a subset pointer to your current position, if the subset pointer is not already set. (Used with DEDBs only.)
Z	Set a subset pointer to 0, so it can be reused. (Used with DEDBs only.)
-	NULL. Use an SSA in command code format without specifying the command code. Can be replaced during execution with the command codes that you want.

Committing and Rolling Back Changes

Let's look at how we commit updated data to the database. This is not difficult to do using checkpoint calls. Using checkpoint **restart** is somewhat more involved, especially for running in DLI mode where you must use a log file. We'll provide examples of both checkpointing and checkpoint restarting. It will be better if we take it in two chunks with two programs, so that's what we'll do.

For COBIMSD our objective is to delete all the records in the database. We use the same walkthrough-the-database code we used in COBIMS3 except we will use GHN to do the walking, and we will add a DLET call after each GHN to delete the root segment. Note: all child segments are automatically deleted when a root segment is deleted. In fact the principle is even broader - all children under a parent segment are deleted if the parent segment is deleted.

We will also set up checkpointing to show its usage. We will need to do four things before checkpointing can work.

1. Change the PSB to include an IO-PCB
2. Add an XRST call before any data related IMS calls are done
3. Add CHKP calls at specified intervals
4. Add code to reset database position after a checkpoint

Modifying the PSB to Add An IO-PCB

We have to back up a bit to make a fundamental change to our PSB. In order to issue IMS service commands like CHKP (as opposed to database retrieval or update commands) you must use a special PCB called the IO-PCB. Programs that run in BMP mode are always defined to use an IO-PCB, but those that run in DLI mode by default do not have to use an IO-PCB (unless they are doing IMS service calls).

Since we have only been running in DLI mode and not issuing IMS service calls, we didn't define our PSB to include an IOPCB. Since we must now use an IO-PCB to use CHKP calls, let's modify our PSB accordingly. The change is very simple and involves adding a **CMPAT=Y** clause after the PSBNAME= clause. Let's create a separate PSB named EMPPSBZ. It will be a clone of the EMPPSB except for the CMPAT=Y. Here is the code:

```
PRINT NOGEN
PCB    TYPE=DB,NAME=EMPLOYEE,KEYLEN=20,PROCOPT=AP
SENSEG NAME=EMPLOYEE,PARENT=0
SENSEG NAME=EMPPAY,PARENT=EMPLOYEE
SENSEG NAME=EMPPAYHS,PARENT=EMPPAY
SENSEG NAME=EMPDEP,PARENT=EMPLOYEE
PSBGEN LANG=COBOL,PSBNAME=EMPLOYEE,CMPAT=YES
END
```

Let's save this as member EMPPSBZ in our library and run the PSBGEN process.

So what practical effect does this have if we use the EMPPSBZ PSB to run a program? Basically this PSB **implicitly** includes an IO-PCB, meaning you don't see an IO-PCB defined in the PSB, but it must be the first PCB pointer in the linkage between your program and IMS. Since we defined the PSB this way, you **must** handle the IO-PCB in your program by:

- Including a structure for the IO-PCB.
- Including the IO-PCB structure name in the ENTRY statement in the procedure division.

Here is our new IO-PCB structure:

```
01 IO-PCB.
   05 FILLER           PICTURE X(10).
   05 IO-STATUS-CODE   PICTURE XX.
   05 FILLER           PICTURE X(20).
```

And here is the change to the ENTRY coded in the procedure division. Notice it now includes both the IO-PCB and the PCB-MASK structures.

```
ENTRY 'DLITCBL' USING IO-PCB, PCB-MASK
```

You MUST put the IO-PCB first in the parameter list before any database PCBs. The database PCBs that follow should be in the same order that they are defined in the PSB. Now we can move on to doing the restart call.

Adding an XRST Call to Initialization Routine

Now we need to include an XRST (Extended Restart Facility) call to check for restart. Don't worry that we won't actually be restarting with this program yet (the reason is because we aren't logging our changes yet – be patient, we'll get there in the next program). The XRST call is part of the procedure that we need to do symbolic checkpoints and eventually perform IMS restarts, so we include it here. [5]

First, add these structures and variables to your working storage section.

```
01 XRST-IOAREA.
   05 XRST-ID        PIC X(08) VALUE SPACES.
   05 FILLER         PIC X(04) VALUE SPACES.

77 IO-AREALEN        PIC S9(9) USAGE IS BINARY VALUE 12.

77 CHKP-ID           PIC X(08) VALUE 'IMSD    '.

77 CHKP-NBR          PIC 999   VALUE ZERO.
77 CHKP-COUNT        PIC S9(9) USAGE IS BINARY VALUE ZERO.

01 CHKP-MESSAGE.
   05 FILLER                  PIC X(24) VALUE
      'COBIMSD  CHECK POINT NO:'.
   05 CHKP-MESS-NBR           PIC 999      VALUE ZERO.
   05 FILLER                  PIC X(15)    VALUE ',AT INPUT REC#:'.

   05 CHKP-MESS-REC           PIC ZZZZZ9   VALUE SPACES.
   05 FILLER                  PIC X(10)    VALUE ',AT EMP#:'.
   05 CHKP-MESS-EMP           PIC X(08)    VALUE SPACES.
```

---

5 In this text we will only deal with symbolic checkpoints. IMS also offers basic checkpoints, but these do not work with the extended restart facility (the XRST call and automated repositions, etc), so with basic checkpoints your program must do 100% of the code to perform a restart. Consequently basic checkpoints are of limited value and I don't deal with them in this text.

```
01  IMS-CHKP-AREA-LTH.
    05 LEN                   PIC S9(9) USAGE IS BINARY VALUE +7.

01  IMS-CHKP-AREA.
    05 CHKP-EMP-ID    PIC X(04) VALUE SPACES.
    05 CHKP-NBR-LAST  PIC 999   VALUE 0.
```

Second, add this code at the beginning of your Initialization paragraph.

```
*  CHECK FOR RESTART

    CALL 'CBLTDLI' USING SIX,
         DLI-FUNCXRST,
         PCB-MASK,
         IO-AREALEN,
         XRST-IOAREA,
         IMS-CHKP-AREA-LTH,
         IMS-CHKP-AREA

    IF STATUS-CODE NOT EQUAL SPACES THEN
       PERFORM P9000-DISPLAY-ERROR
       GOBACK
    END-IF

    IF XRST-ID NOT EQUAL SPACES THEN
       MOVE CHKP-NBR-LAST TO CHKP-NBR
       DISPLAY '*** COBIMSD IMS RESTART ***'
       DISPLAY '*  LAST CHECK POINT :' XRST-ID
       DISPLAY '*  EMPLOYEE NUMBER  :' CHKP-EMP-ID
    ELSE
       DISPLAY '****** COBIMSD IMS NORMAL START ***'
       PERFORM P8000-TAKE-CHECKPOINT
    END-IF.
```

This code checks to see if our execution is being run as a restart. If it is, then we announce that it is a restart. If it is not, we announce a normal start. That's all we need to do with XRST right now. Later we will add code to perform the various restart actions, and we'll explain the parameters at that time.

Adding the CHKP Call

Now let's add code for taking a checkpoint. We'll code a separate procedure for this. The required parameters for the call are the CHKP function, the IO-PCB structure, the length of an IO area that contains the checkpoint id, the IO area that contains the checkpoint id, the length of the checkpoint area, and the checkpoint area structure.

The latter is where you save anything you want to save for restart, such as the last processed EMP-ID, record counters and anything else you want to save for a restart.

Here is the code for doing the checkpoint call.

```
P8000-TAKE-CHECKPOINT.
      DISPLAY 'PROCESSING IN P8000-TAKE-CHECKPOINT'
      ADD +1              TO CHKP-NBR
      MOVE CHKP-NBR       TO CHKP-NBR-LAST
      MOVE CHKP-NBR-LAST  TO CHKP-ID(6:3)
      MOVE EMP-ID         TO CHKP-EMP-ID

      CALL 'CBLTDLI' USING SIX,
            DLI-FUNCCHKP,
            IO-PCB,
            IO-AREALEN,
            CHKP-ID,
            IMS-CHKP-AREA-LTH,
            IMS-CHKP-AREA

      IF IO-STATUS-CODE NOT EQUAL SPACES THEN
          DISPLAY 'TOOK AN ERROR DOING THE CHECKPOINT'
          DISPLAY 'IO-STATUS-CODE ' IO-STATUS-CODE
          PERFORM P9000-DISPLAY-ERROR
          PERFORM P9000-DISPLAY-ERROR
          GOBACK
      ELSE
          MOVE 0 TO CHKP-COUNT
          MOVE CHKP-NBR        TO CHKP-MESS-NBR
          MOVE CHKP-EMP-ID     TO CHKP-MESS-EMP
          DISPLAY CHKP-MESSAGE
      END-IF.
```

One final note: the third parameter in the CHKP call (the IO area length) is not actually used by IMS, but it must still be included for backward compatibility. You need only define a variable for it in the program.

Adding Code to Reposition in the Database After Checkpoint

Finally, we must create code to reposition the database after taking a checkpoint. The reason is that the checkpoint call causes the database position to be lost. If you continue GHN calls at this point without reestablishing your database position, you'll get an error.

So what we'll do is to ensure we have the next record to process and we'll include that in the checkpoint IO area that we are going to save. So our code will:

DLET a record
Read the next record and capture the employee id
If it is time to take a checkpoint then
       Take a check point using the captured employee id that was just read
       Reposition in the database using the captured employee id

The reposition code is as follows. Notice it is using a qualified SSA to get the exact record that is needed to reposition. Of course we must use a qualified SSA, and the EMP-ID that was retrieved in the GHN call before we took the checkpoint.

```
P1000-RESET-POSITION.

    DISPLAY 'PROCESSING IN P1000-RESET-POSITION'

    CALL 'CBLTDLI' USING FOUR,
        DLI-FUNCGHU,
        PCB-MASK,
        IO-EMPLOYEE-RECORD,
        EMP-QUALIFIED-SSA

    IF STATUS-CODE NOT EQUAL SPACES THEN
        PERFORM P9000-DISPLAY-ERROR
        GOBACK
    ELSE
        DISPLAY 'SUCCESSFUL REPOSITION AT EMP ID ' EMP-ID.
```

Ok, now we've performed all four items that will enable us to commit data updates by taking checkpoints at some interval. Let's make our record interval 5. So we have eight records in the database, and we'll take a checkpoints as follows:

- At the beginning of the program.
- After each 5 records have been processed.
- At the end of the program.

Here is our complete program code for COBIMSD. As mentioned earlier, we haven't completed the code yet for a restart. But we now have the functionality to commit our data changes with the checkpoint call.

```
IDENTIFICATION DIVISION.
PROGRAM-ID. COBIMSD.
**********************************************************
*   WALK THROUGH THE EMPLOYEE (ROOT) SEGMENTS OF       *
*   THE ENTIRE EMPLOYEE DATABASE. DELETE ALL RECORDS.*
**********************************************************
ENVIRONMENT DIVISION.
INPUT-OUTPUT SECTION.
DATA DIVISION.

**********************************************************
*  W O R K I N G    S T O R A G E    S E C T I O N    *
**********************************************************
WORKING-STORAGE SECTION.

 01 WS-FLAGS.
     05   SW-END-OF-DB-SWITCH     PIC X(1) VALUE 'N'.
         88   SW-END-OF-DB                  VALUE 'Y'.
         88   SW-NOT-END-OF-DB              VALUE 'N'.

 01 DLI-FUNCTIONS.
     05 DLI-FUNCISRT  PIC X(4) VALUE 'ISRT'.
     05 DLI-FUNCGU    PIC X(4) VALUE 'GU  '.
     05 DLI-FUNCGN    PIC X(4) VALUE 'GN  '.
     05 DLI-FUNCGHU   PIC X(4) VALUE 'GHU '.
     05 DLI-FUNCGHN   PIC X(4) VALUE 'GHN '.
     05 DLI-FUNCGNP   PIC X(4) VALUE 'GNP '.
     05 DLI-FUNCREPL  PIC X(4) VALUE 'REPL'.
     05 DLI-FUNCDLET  PIC X(4) VALUE 'DLET'.
     05 DLI-FUNCXRST  PIC X(4) VALUE 'XRST'.
     05 DLI-FUNCCHKP  PIC X(4) VALUE 'CHKP'.

 01 IO-EMPLOYEE-RECORD.
     05   EMP-ID        PIC X(04).
     05   FILLER        PIC X(01).
     05   EMPL-LNAME    PIC X(30).
     05   FILLER        PIC X(01).
     05   EMPL-FNAME    PIC X(20).
     05   FILLER        PIC X(01).
     05   EMPL-YRS-SRV  PIC X(02).
     05   FILLER        PIC X(01).
     05   EMPL-PRM-DTE  PIC X(10).
     05   FILLER        PIC X(10).

 01 EMP-UNQUALIFIED-SSA.
     05   SEGNAME       PIC X(08) VALUE 'EMPLOYEE'.
     05   FILLER        PIC X(01) VALUE ' '.
```

```
   01 EMP-QUALIFIED-SSA.
      05  SEGNAME     PIC X(08) VALUE 'EMPLOYEE'.
      05  FILLER      PIC X(01) VALUE '('.
      05  FIELD       PIC X(08) VALUE 'EMPID'.
      05  OPER        PIC X(02) VALUE ' ='.
      05  EMP-ID-VAL  PIC X(04) VALUE '    '.
      05  FILLER      PIC X(01) VALUE ')'.

   01 SEG-IO-AREA    PIC X(80).

 01 IMS-RET-CODES.
      05 ONE          PIC S9(9) COMP VALUE +1.
      05 TWO          PIC S9(9) COMP VALUE +2.
      05 THREE        PIC S9(9) COMP VALUE +3.
      05 FOUR         PIC S9(9) COMP VALUE +4.
      05 FIVE         PIC S9(9) COMP VALUE +5.
      05 SIX          PIC S9(9) COMP VALUE +6.


   01 XRST-IOAREA.
      05 XRST-ID      PIC X(08) VALUE SPACES.
      05 FILLER       PIC X(04) VALUE SPACES.

   77 IO-AREALEN     PIC S9(9) USAGE IS BINARY VALUE 12.

   77 CHKP-ID        PIC X(08) VALUE 'IMSD     '.

   77 CHKP-NBR       PIC 999   VALUE ZERO.
   77 CHKP-COUNT     PIC S9(9) USAGE IS BINARY VALUE ZERO.

   01 CHKP-MESSAGE.
      05 FILLER                PIC X(24) VALUE
         'COBIMSD  CHECK POINT NO:'.
      05 CHKP-MESS-NBR         PIC 999      VALUE ZERO.
      05 FILLER                PIC X(15)    VALUE ',AT INPUT REC#:'.
      05 CHKP-MESS-REC         PIC ZZZZZ9   VALUE SPACES.
      05 FILLER                PIC X(10)    VALUE ',AT EMP#:'.
      05 CHKP-MESS-EMP         PIC X(08)    VALUE SPACES.

   01 IMS-CHKP-AREA-LTH.
      05 LEN                PIC S9(9) USAGE IS BINARY VALUE +7.

   01 IMS-CHKP-AREA.
      05 CHKP-EMP-ID     PIC X(04) VALUE SPACES.
      05 CHKP-NBR-LAST   PIC 999   VALUE 0.

LINKAGE SECTION.
```

```cobol
01 IO-PCB.
   05 FILLER          PICTURE X(10).
   05 IO-STATUS-CODE  PICTURE XX.
   05 FILLER          PICTURE X(20).

01 PCB-MASK.
   03 DBD-NAME        PIC X(8).
   03 SEG-LEVEL       PIC XX.
   03 STATUS-CODE     PIC XX.
   03 PROC-OPT        PIC X(4).
   03 FILLER          PIC X(4).
   03 SEG-NAME        PIC X(8).
   03 KEY-FDBK        PIC S9(5) COMP.
   03 NUM-SENSEG      PIC S9(5) COMP.
   03 KEY-FDBK-AREA.
      05 EMPLOYEE-KEY  PIC X(04).
      05 EMPPAYHS-KEY  PIC X(08).

PROCEDURE DIVISION.

    INITIALIZE IO-PCB PCB-MASK
    ENTRY 'DLITCBL' USING IO-PCB, PCB-MASK

    PERFORM P100-INITIALIZATION.
    PERFORM P200-MAINLINE.
    PERFORM P300-TERMINATION.
    GOBACK.

P100-INITIALIZATION.

    DISPLAY '** PROGRAM COBIMSD START **'
    DISPLAY 'PROCESSING IN P100-INITIALIZATION'.

* CHECK FOR RESTART

    CALL 'CBLTDLI' USING SIX,
         DLI-FUNCXRST,
         PCB-MASK,
         IO-AREALEN,
         XRST-IOAREA,
         IMS-CHKP-AREA-LTH,
         IMS-CHKP-AREA

    IF STATUS-CODE NOT EQUAL SPACES THEN
       PERFORM P9000-DISPLAY-ERROR
       GOBACK
    END-IF
```

```
        IF XRST-ID NOT EQUAL SPACES THEN
           MOVE CHKP-NBR-LAST TO CHKP-NBR
           DISPLAY '*** COBIMSD IMS RESTART ***'
           DISPLAY '*  LAST CHECK POINT :' XRST-ID
           DISPLAY '*  EMPLOYEE NUMBER  :' CHKP-EMP-ID
        ELSE
           DISPLAY '****** COBIMSD IMS NORMAL START ***'
           PERFORM P8000-TAKE-CHECKPOINT
        END-IF.

*    DO INITIAL DB READ FOR FIRST EMPLOYEE RECORD

        CALL 'CBLTDLI' USING FOUR,
             DLI-FUNCGHN,
             PCB-MASK,
             IO-EMPLOYEE-RECORD,
             EMP-UNQUALIFIED-SSA

        IF STATUS-CODE = '  ' THEN
           NEXT SENTENCE
        ELSE
           IF STATUS-CODE = 'GB' THEN
              SET SW-END-OF-DB TO TRUE
              DISPLAY 'END OF DATABASE :'
           ELSE
              PERFORM P9000-DISPLAY-ERROR
              GOBACK
           END-IF

        END-IF.

   P200-MAINLINE.

        DISPLAY 'PROCESSING IN P200-MAINLINE'

*    CHECK STATUS CODE AND FIRST RECORD

        IF SW-END-OF-DB THEN
           DISPLAY 'NO RECORDS TO PROCESS!!'
        ELSE

           PERFORM UNTIL SW-END-OF-DB

              CALL 'CBLTDLI' USING THREE,
                   DLI-FUNCDLET,
                   PCB-MASK,
                   IO-EMPLOYEE-RECORD
```

161

```
            IF STATUS-CODE NOT EQUAL SPACES THEN
                PERFORM P9000-DISPLAY-ERROR
                GOBACK
            ELSE
                DISPLAY 'SUCCESSFUL DELETE OF EMPLOYEE ' EMP-ID
            END-IF

*       GET THE NEXT RECORD

            CALL 'CBLTDLI' USING FOUR,
                DLI-FUNCGHN,
                PCB-MASK,
                IO-EMPLOYEE-RECORD,
                EMP-UNQUALIFIED-SSA

            IF STATUS-CODE = 'GB' THEN
                SET SW-END-OF-DB TO TRUE
                DISPLAY 'END OF DATABASE'
            ELSE
                IF STATUS-CODE NOT EQUAL SPACES THEN
                    PERFORM P9000-DISPLAY-ERROR
                    SET SW-END-OF-DB TO TRUE
                    GOBACK
                ELSE
                    DISPLAY 'SUCCESSFUL GET HOLD :'
                        IO-EMPLOYEE-RECORD
                    MOVE EMP-ID TO EMP-ID-VAL
                    ADD +1 TO CHKP-COUNT
                    IF CHKP-COUNT GREATER THAN OR EQUAL TO 5
                        PERFORM P8000-TAKE-CHECKPOINT
                        PERFORM P1000-RESET-POSITION
                    END-IF
                END-IF
            END-IF

    END-PERFORM.
    DISPLAY 'FINISHED PROCESSING IN P200-MAINLINE'.

P300-TERMINATION.

    DISPLAY 'PROCESSING IN P300-TERMINATION'
    ADD +1 TO CHKP-COUNT
    PERFORM P8000-TAKE-CHECKPOINT
    DISPLAY '** COBIMSD - SUCCESSFULLY ENDED **'.

P1000-RESET-POSITION.
```

```
          DISPLAY 'PROCESSING IN P1000-RESET-POSITION'

          CALL 'CBLTDLI' USING FOUR,
                DLI-FUNCGHU,
                PCB-MASK,
                IO-EMPLOYEE-RECORD,
                EMP-QUALIFIED-SSA

          IF STATUS-CODE NOT EQUAL SPACES THEN
             PERFORM P9000-DISPLAY-ERROR
             GOBACK
          ELSE
             DISPLAY 'SUCCESSFUL REPOSITION AT EMP ID ' EMP-ID.

     P8000-TAKE-CHECKPOINT.

          DISPLAY 'PROCESSING IN P8000-TAKE-CHECKPOINT'

          ADD +1             TO CHKP-NBR
          MOVE CHKP-NBR       TO CHKP-NBR-LAST
          MOVE CHKP-NBR-LAST TO CHKP-ID(6:3)
          MOVE EMP-ID         TO CHKP-EMP-ID

          CALL 'CBLTDLI' USING SIX,
                DLI-FUNCCHKP,
                IO-PCB,
                IO-AREALEN,
                CHKP-ID,
                IMS-CHKP-AREA-LTH,
                IMS-CHKP-AREA

          IF IO-STATUS-CODE NOT EQUAL SPACES THEN
             DISPLAY 'TOOK AN ERROR DOING THE CHECKPOINT'
             DISPLAY 'IO-STATUS-CODE ' IO-STATUS-CODE
             PERFORM P9000-DISPLAY-ERROR
             GOBACK
          ELSE
             MOVE 0 TO CHKP-COUNT
             MOVE CHKP-NBR        TO CHKP-MESS-NBR
             MOVE CHKP-EMP-ID     TO CHKP-MESS-EMP
             DISPLAY CHKP-MESSAGE
          END-IF.

     P9000-DISPLAY-ERROR.

          DISPLAY 'ERROR ENCOUNTERED - DETAIL FOLLOWS'
          DISPLAY 'SEG-IO-AREA      :' SEG-IO-AREA
          DISPLAY 'DBD-NAME1:'      DBD-NAME
```

163

```
            DISPLAY 'SEG-LEVEL1:'    SEG-LEVEL
            DISPLAY 'STATUS-CODE:'   STATUS-CODE
            DISPLAY 'PROC-OPT1 :'    PROC-OPT
            DISPLAY 'SEG-NAME1 :'    SEG-NAME
            DISPLAY 'KEY-FDBK1 :'    KEY-FDBK
            DISPLAY 'NUM-SENSEG1:'   NUM-SENSEG
            DISPLAY 'KEY-FDBK-AREA1:' KEY-FDBK-AREA.

    *    END OF SOURCE CODE
```

At this point, we can compile and link, and then run the program. Make sure your JCL specifies the EMPPSBZ PSB or you'll get an error.

```
** PROGRAM COBIMSD START **
PROCESSING IN P100-INITIALIZATION
****** COBIMSD IMS NORMAL START ***
PROCESSING IN P8000-TAKE-CHECKPOINT
COBIMSD   CHECK POINT NO:001,AT INPUT REC#:       ,AT EMP#:
PROCESSING IN P200-MAINLINE
SUCCESSFUL DELETE OF EMPLOYEE 1111
SUCCESSFUL GET HOLD :1122 JENKINS                    DEBORAH           05
SUCCESSFUL DELETE OF EMPLOYEE 1122
SUCCESSFUL GET HOLD :3217 JOHNSON                    EDWARD            04
SUCCESSFUL DELETE OF EMPLOYEE 3217
SUCCESSFUL GET HOLD :4175 TURNBULL                   FRED              01
SUCCESSFUL DELETE OF EMPLOYEE 4175
SUCCESSFUL GET HOLD :4720 SCHULTZ                    TIM               09
SUCCESSFUL DELETE OF EMPLOYEE 4720
SUCCESSFUL GET HOLD :4836 SMITH                      SANDRA            03
PROCESSING IN P8000-TAKE-CHECKPOINT
COBIMSD   CHECK POINT NO:002,AT INPUT REC#:       ,AT EMP#: 4836
PROCESSING IN P1000-RESET-POSITION
SUCCESSFUL REPOSITION AT EMP ID 4836
SUCCESSFUL DELETE OF EMPLOYEE 4836
SUCCESSFUL GET HOLD :6288 WILLARD                    JOE               06
SUCCESSFUL DELETE OF EMPLOYEE 6288
SUCCESSFUL GET HOLD :7459 STEWART                    BETTY             07
SUCCESSFUL DELETE OF EMPLOYEE 7459
END OF DATABASE
FINISHED PROCESSING IN P200-MAINLINE
PROCESSING IN P300-TERMINATION
PROCESSING IN P8000-TAKE-CHECKPOINT
COBIMSD   CHECK POINT NO:003,AT INPUT REC#:       ,AT EMP#: 7459
** COBIMSD - SUCCESSFULLY ENDED **
```

We now have an empty database. You can verify this by looking in your File Manager IMS if you have it, or you can try browsing the DATA file of the KSDS. Since it is empty, you'll get an error.

```
VSAM POINT RC X"08", Error Code X"20"
VSAM GET RC X"08", Error Code X"58"
Function terminated
***
```

We have shown we can commit updates to the database at some interval. In a real production environment we would not checkpoint every 5 records. More likely we would checkpoint at 500 records or 1,000 records or 2,000 records. You don't want to lock your data for too long, so find a record interval that commits at about once a minute, or whatever your DBA recommends.

Here is the corresponding PLI code for COBIMSD.

```
PLIIMSD: PROCEDURE (IO_PTR_PCB,DB_PTR_PCB) OPTIONS(MAIN);
/*******************************************************************
* PROGRAM NAME: PLIIMSD - WALK THROUGH THE EMPLOYEE (ROOT)        *
*                         SEGMENTS OF THE EMPLOYEE IMS DATABASE,   *
*                         AND DELETE EACH ONE.                     *
*******************************************************************/


/*******************************************************************
/*               W O R K I N G   S T O R A G E                     *
*******************************************************************/

    DCL SW_END_OF_DB            STATIC BIT(01) INIT('0'B);
    DCL ONCODE                  BUILTIN;
    DCL DB_PTR_PCB              POINTER;
    DCL IO_PTR_PCB              POINTER;

    DCL PLITDLI                 EXTERNAL ENTRY;

    DCL 01 DLI_FUNCTIONS,
           05 DLI_FUNCISRT      CHAR(04) INIT ('ISRT'),
           05 DLI_FUNCGU        CHAR(04) INIT ('GU  '),
           05 DLI_FUNCGN        CHAR(04) INIT ('GN  '),
           05 DLI_FUNCGHU       CHAR(04) INIT ('GHU '),
           05 DLI_FUNCGNP       CHAR(04) INIT ('GNP '),
           05 DLI_FUNCREPL      CHAR(04) INIT ('REPL'),
           05 DLI_FUNCDLET      CHAR(04) INIT ('DLET'),
           05 DLI_FUNCXRST      CHAR(04) INIT ('XRST'),
           05 DLI_FUNCCHKP      CHAR(04) INIT ('CHKP'),
           05 DLI_FUNCROLL      CHAR(04) INIT ('ROLL');

    DCL 01 IO_EMPLOYEE_RECORD,
           05  EMPL_ID          CHAR(04),
           05  FILLER1          CHAR(01),
```

165

```
               05   EMPL_LNAME          CHAR(30),
               05   FILLER2             CHAR(01),
               05   EMPL_FNAME          CHAR(20),
               05   FILLER3             CHAR(01),
               05   EMPL_YRS_SRV        CHAR(02),
               05   FILLER4             CHAR(01),
               05   EMPL_PRM_DTE        CHAR(10),
               05   FILLER5             CHAR(10);

     DCL 01 EMP_UNQUALIFIED_SSA,
               05   SEGNAME             CHAR(08) INIT ('EMPLOYEE'),
               05   FILLER7             CHAR(01) INIT (' ');

     DCL 01 EMP_QUALIFIED_SSA,
               05   SEGNAME             CHAR(08) INIT('EMPLOYEE'),
               05   FILLER8             CHAR(01) INIT('('),
               05   FIELD               CHAR(08) INIT('EMPID'),
               05   OPER                CHAR(02) INIT(' ='),
               05   EMP_ID_VAL          CHAR(04) INIT('    '),
               05   FILLER9             CHAR(01) INIT(')');

     DCL THREE                          FIXED BIN (31) INIT(3);
     DCL FOUR                           FIXED BIN (31) INIT(4);
     DCL FIVE                           FIXED BIN (31) INIT(5);
     DCL SIX                            FIXED BIN (31) INIT(6);

     DCL 01 XRST_IOAREA,
          05 XRST_ID      CHAR(08) INIT('        '),
          05 FILLER10     CHAR(04) INIT('     ');

     DCL XRST_IO_AREALEN    FIXED BIN(31) INIT (12);
     DCL IO_AREALEN         FIXED BIN(31) INIT (08);
     DCL CHKP_ID            CHAR(08) INIT( 'IMSD-   ');
     DCL CHKP_NBR           FIXED DEC (3)  INIT(0);
     DCL CHKP_COUNT         FIXED BIN (31) INIT(0);

     DCL 01 CHKP_MESSAGE,
          05 FILLER11          CHAR(24)    INIT(
             'COBIMSD  CHECK POINT NO:'),
          05 CHKP_MESS_NBR   PIC '999',
          05 FILLER12          CHAR(15)    INIT( ',AT INPUT REC#:'),
          05 CHKP_MESS_REC   PIC 'ZZZZZ9',
          05 FILLER13          CHAR(10)    INIT(',AT EMP#:'),
          05 CHKP_MESS_EMP   CHAR(08)    INIT(' ');

     DCL IMS_CHKP_AREA_LTH    FIXED BIN (31) INIT(07);

     DCL 01 IMS_CHKP_AREA,
          05 CHKP_EMP_ID    CHAR(04)    INIT('0000'),
          05 CHKP_NBR_LAST  CHAR(03)    INIT('000');

     DCL 01 IO_PCB                  BASED(IO_PTR_PCB),
          05 FILLER97         CHAR(10)  INIT(' '),
          05 IO_STATUS_CODE   CHAR(02)  INIT (' ');
```

```
DCL 01 PCB_MASK                BASED(DB_PTR_PCB),
        05 DBD_NAME            CHAR(08),
        05 SEG_LEVEL           CHAR(02),
        05 STATUS_CODE         CHAR(02),
        05 PROC_OPT            CHAR(04),
        05 FILLER99            FIXED BIN (31),
        05 SEG_NAME            CHAR(08),
        05 KEY_FDBK            FIXED BIN (31),
        05 NUM_SENSEG          FIXED BIN (31),
        05 KEY_FDBK_AREA,
          10 KFB_EMPLOYEE_ID CHAR(04);

/**********************************************************************
/*                 P R O G R A M   M A I N L I N E                   *
**********************************************************************/

CALL P100_INITIALIZATION;
CALL P200_MAINLINE;
CALL P300_TERMINATION;

P100_INITIALIZATION: PROC;

    PUT SKIP LIST ('PLIIMSD: TRAVERSE EMPLOYEE DATABASE ROOT SEGS');
    PUT SKIP LIST ('PROCESSING IN P100_INITIALIZATION');
    IO_PCB   = '';
    PCB_MASK = '';
    IO_EMPLOYEE_RECORD  = '';

    /* CHECK FOR RESTART */

    CALL PLITDLI (SIX,
                  DLI_FUNCXRST,
                  IO_PCB,
                  XRST_IO_AREALEN,
                  XRST_IOAREA,
                  IMS_CHKP_AREA_LTH,
                  IMS_CHKP_AREA);

    IF IO_STATUS_CODE ¬= '  ' THEN
        DO;
           CALL P9000_DISPLAY_ERROR;
           RETURN;
        END;

    IF XRST_ID ¬= ' ' THEN
        DO;
           CHKP_NBR = CHKP_NBR_LAST;
           PUT SKIP LIST ('*** COBIMSD IMS RESTART ***');
           PUT SKIP LIST ('*  LAST CHECK POINT :' || XRST_ID);
           PUT SKIP LIST ('*  EMPLOYEE NUMBER  :' || CHKP_EMP_ID);
        END;
    ELSE
        DO;
```

167

```
                    PUT SKIP LIST ('****** COBIMSD IMS NORMAL START ***');
                    CALL P8000_TAKE_CHECKPOINT;
               END;

     /* DO INITIAL DB READ FOR FIRST EMPLOYEE RECORD */

          CALL PLITDLI (FOUR,
                         DLI_FUNCGHN,
                         PCB_MASK,
                         IO_EMPLOYEE_RECORD,
                         EMP_UNQUALIFIED_SSA);

          IF STATUS_CODE = '  ' THEN;
          ELSE
              IF STATUS_CODE = 'GB' THEN
                  DO;
                      SW_END_OF_DB = '1'B;
                      PUT SKIP LIST ('** END OF DATABASE');
                  END;
              ELSE
                  DO;
                      CALL P9000_DISPLAY_ERROR;
                      RETURN;
                  END;

     END P100_INITIALIZATION;

     P200_MAINLINE: PROC;

         /*  MAIN LOOP - CYCLE THROUGH ALL ROOT SEGMENTS IN THE DB,
                         DISPLAYING THE DATA RETRIEVED              */

         IF SW_END_OF_DB THEN
             PUT SKIP LIST ('NO RECORDS TO PROCESS!!');
         ELSE
             DO WHILE (¬SW_END_OF_DB);
                 PUT SKIP LIST ('SUCCESSFUL GET-HOLD :'
                     || EMPL_ID);

                 /* DELETE THE SWGMENT */

                 CALL PLITDLI (THREE,
                                DLI_FUNCDLET,
                                PCB_MASK,
                                IO_EMPLOYEE_RECORD);

                 IF STATUS_CODE ¬= '  ' THEN
                     DO;
                         CALL P9000_DISPLAY_ERROR;
                         RETURN;
                     END;
                 ELSE
                     PUT SKIP LIST ('SUCCESSFUL DELETE OF EMP ' || EMPL_ID);
```

```
            /* NOW GET THE NEXT ROOT TO DELETE */

        CALL PLITDLI (FOUR,
                      DLI_FUNCGN,
                      PCB_MASK,
                      IO_EMPLOYEE_RECORD,
                      EMP_UNQUALIFIED_SSA);

        IF STATUS_CODE = ' ' THEN
            DO;
                PUT SKIP LIST ('SUCCESSFUL GET HOLD: ' || EMPL_ID);
                EMP_ID_VAL = EMPL_ID;
                CHKP_COUNT = CHKP_COUNT + 1;
                IF CHKP_COUNT >= 5 THEN
                    DO;
                        CALL P8000_TAKE_CHECKPOINT;
                        CALL P1000_RESET_POSITION;
                    END;
            END;
        ELSE
            IF STATUS_CODE = 'GB' THEN
                DO;
                    SW_END_OF_DB = '1'B;
                    PUT SKIP LIST ('** END OF DATABASE');
                END;
            ELSE
                DO;
                    CALL P9000_DISPLAY_ERROR;
                    RETURN;
                END;

    END; /* DO WHILE */

 PUT SKIP LIST ('FINISHED PROCESSING IN P200_MAINLINE');

END P200_MAINLINE;

P300_TERMINATION: PROC;

 PUT SKIP LIST ('PROCESSING IN P300_TERMINATION');

 CHKP_COUNT = CHKP_COUNT + 1;
 CALL P8000_TAKE_CHECKPOINT;

 PUT SKIP LIST ('PLIIMSD - SUCCESSFULLY ENDED');

END P300_TERMINATION;

P1000_RESET_POSITION: PROC;

 PUT SKIP LIST ('PROCESSING IN P1000_RESET_POSITION');

 CALL PLITDLI (FOUR,
               DLI_FUNCGHU,
```

```
                        PCB_MASK,
                        IO_EMPLOYEE_RECORD,
                        EMP_QUALIFIED_SSA);

    IF STATUS_CODE ¬= ' ' THEN
        DO;
            CALL P9000_DISPLAY_ERROR;
            RETURN;
        END;
    ELSE
        PUT SKIP LIST ('SUCCESSFUL REPOSITION AT EMP ID ' || EMPL_ID);

END P1000_RESET_POSITION;

P8000_TAKE_CHECKPOINT: PROC;

    PUT SKIP LIST ('PROCESSING IN P8000_TAKE_CHECKPOINT');

    CHKP_NBR            = CHKP_NBR + 1;
    CHKP_NBR_LAST       = CHKP_NBR;
    SUBSTR(CHKP_ID,6,3) = CHKP_NBR_LAST;
    CHKP_EMP_ID         = EMPL_ID;

    PUT SKIP LIST ('IO_AREALEN ' || IO_AREALEN);
    PUT SKIP LIST ('IMS_CHKP_AREA_LTH ' || IMS_CHKP_AREA_LTH);

    PUT SKIP LIST ('CHKP_ID = ' || CHKP_ID);
    PUT SKIP LIST (' ');

    CALL PLITDLI (SIX,
                  DLI_FUNCCHKP,
                  IO_PCB,
                  IO_AREALEN,
                  CHKP_ID,
                  IMS_CHKP_AREA_LTH,
                  IMS_CHKP_AREA);

    IF IO_STATUS_CODE ¬= ' ' THEN
        DO;
            PUT SKIP LIST ('TOOK AN ERROR DOING THE CHECKPOINT');
            PUT SKIP LIST ('IO_STATUS_CODE ' || IO_STATUS_CODE);
            CALL P9000_DISPLAY_ERROR;
            RETURN;
        END;
    ELSE
        DO;
            CHKP_COUNT = 0;
            CHKP_MESS_NBR = CHKP_NBR;
            CHKP_MESS_EMP = CHKP_EMP_ID;
            PUT SKIP LIST (CHKP_MESSAGE);
        END;

END P8000_TAKE_CHECKPOINT;
```

```
P9000_DISPLAY_ERROR: PROC;

    PUT SKIP LIST ('ERROR ENCOUNTERED - DETAIL FOLLOWS');
    PUT SKIP LIST ('SEG_IO_AREA      :' || SEG_IO_AREA);
    PUT SKIP LIST ('DBD_NAME1:' || DBD_NAME);
    PUT SKIP LIST ('SEG_LEVEL1:' || SEG_LEVEL);
    PUT SKIP LIST ('STATUS_CODE:' || STATUS_CODE);
    PUT SKIP LIST ('PROC_OPT1 :' || PROC_OPT);
    PUT SKIP LIST ('SEG_NAME1 :' || SEG_NAME);
    PUT SKIP LIST ('KEY_FDBK1 :' || KEY_FDBK);
    PUT SKIP LIST ('NUM_SENSEG1:' || NUM_SENSEG);
    PUT SKIP LIST ('KEY_FDBK_AREA1:' || KEY_FDBK_AREA);

END P9000_DISPLAY_ERROR;

END PLIIMSD;
```

Performing Checkpoint Restart

At this point, we've successfully committed data using checkpoints. However, we have not yet demonstrated how to perform a restart using the extended restart facility (XRST). To do that, we need to introduce IMS logging.

Using the IMS Log

To allow for IMS restartability, you must log all the transactions and checkpoints you take. When you stop the program (or when IMS stops it for an abend), your data modifications (ISRT, REPL, DLET) are automatically backed out to the last checkpoint. So typically, you will want to fix whatever the problem was, and then restart your program from the last checkpoint.

In your execution JCL for running IMS programs, there should be two DD statements that are probably dummied out.[6] The IEFRDER DD should definitely be there, and the IMSLOGR may be there (it is only referenced on restart so it might not be).

```
//IMSLOGR   DD DUMMY
//IEFRDER   DD DUMMY
```

---

6 This discussion pertains to running a program in DLI mode. If you are running a program in BMP mode, you don't need these DDs because the program runs in the IMS online space which has its own transaction log.

Here's what these are used for when they are not dummied out (when actual file names are specified):

IMSLOGR – the previous (existing) generation of IMS log file created for your DLI execution.

IEFRDER – the new generation of the IMS log file created for your DLI execution to log any updates to the database performed by your program.

You'll want to create a generation data group for your IMS log file, and then define these DDs to use the 0 and +1 generation of this data set. I created USER01.IMSLOG with 5 generations, and I created an empty first generation. Next, I have un-dummied the IMSLOGR and IEFRDER DD's by coding the new log file as follows:

```
//IMSLOGR   DD DSN=USER01.IMSLOG(+0),
//          DISP=SHR
//IEFRDER   DD DSN=USER01.IMSLOG(+1),
//          DISP=(NEW,CATLG,CATLG),
//          UNIT=SYSDA,
//          SPACE=(TRK,(1,1),RLSE),
//          DCB=(RECFM=VB,BLKSIZE=4096,
//          LRECL=4092,BUFNO=2)
```

Now if you specify a checkpoint value when you restart your program, IMS will scan the 0 generation of the IMS log to pick up the information from the last checkpoint. In our case, this information includes the employee id that we read before issuing the last checkpoint. You can then use that employee id key to reposition in the database.

Specifying a Checkpoint ID on Restart

You can specify the checkpoint id in the PARM value of the execute statement for your program. This is a positional parameter, so it must be placed correctly in the PARM sequence. Here is the JCL and I'm putting a sample checkpoint id at the right place in the PARM.

```
//GO       EXEC PGM=DFSRRC00,REGION=4M,
//      PARM=(DLI,&MBR,&PSB,7,0000,,0,'CHKP0003',N,0,0,,,N,N,,N,)
```

Restart Example

We need to reload the database now before we can do a restart example (remember we deleted all the records in the database earlier). You can run your COBIMS1 to do this. Although the database is empty, it is not brand new. So you can use PSB EMPPSB instead of EMPPSBL. In fact you'll get an error (AI status code) if you use the EMPPSBL, so make sure you use EMPPSB.

When finished, verify that we have nine records in the database.

```
Browse             USER01.IMS.EMPLOYEE.DATA                 Top of 9
Command ===>                                                Scroll PAGE
                        Type DATA     RBA                   Format CHAR
                                      Col 1
----+----10---+----2----+----3----+----4----+----5----+----6----+----7----+----
****  Top of data   ****
......1111 VEREEN                 CHARLES         12 2017-01-01 93
......1122 JENKINS                DEBORAH         05 2017-01-01 43
......3217 JOHNSON                EDWARD          04 2017-01-01 39
......4175 TURNBULL               FRED            01 2016-12-01 54
......4720 SCHULTZ                TIM             09 2017-01-01 65
......4836 SMITH                  SANDRA          03 2017-01-01 02
......6288 WILLARD                JOE             06 2016-01-01 20
......7459 STEWART                BETTY           07 2016-07-31 01
......9134 FRANKLIN               BRIANNA         00 2016-10-01 93
****  End of data   ****
```

For our example, we will create a new program COBIMSE and it will delete all the records in the database as we did with COBIMSD. We will checkpoint at 5 record intervals. You can start by copying COBIMSD to create COBIMSE. There will be two differences between COBIMSD and COBIMSE. One is that COBIMSE will intentionally cause a rollback when we encounter employee 7459 (this is just to simulate an abend type error). The rollback will back out all changes made since the last checkpoint.

The other difference is that we will code restart logic in COBIMSE to reposition to the appropriate employee id in the data to continue processing on a restart. In between run 1 and run 2 of COBIMSE, the only change we will make to the program is to not do the rollback when it gets to employee id 7459. We're simulating a "problem" to cause the rollback, then we solve the cause of the rollback and restart the program.

If you copy COBIMSD to create COBIMSE, you only need to make a few changes. First, let's create some new procedures. One procedure will get the first root in the database. We've been doing that in P100-INITIALIZATION, but now on a restart we need to call the reset position procedure instead. Separating these functions into separate procedures makes the code easier to read. Let's do this:

```
P1000-GET-FIRST-ROOT.

    CALL 'CBLTDLI' USING ,
         DLI-FUNCGHN,
         PCB-MASK,
         IO-EMPLOYEE-RECORD,
         EMP-UNQUALIFIED-SSA

    IF STATUS-CODE = ' ' THEN
       NEXT SENTENCE
    ELSE
       IF STATUS-CODE = 'GB' THEN
          SET SW-END-OF-DB TO TRUE
          DISPLAY 'END OF DATABASE :'
       ELSE
          PERFORM P9000-DISPLAY-ERROR
          GOBACK
       END-IF.
```

Next let's rename P1000-RESET-POSITION to P2000-RESET-POSITION. That will keep the code more orderly. Finally, let's add the procedure to perform the rollback.

```
P3000-ROLLBACK.

    DISPLAY 'PROCESSING IN P3000-ROLLBACK'.

    CALL 'CBLTDLI' USING ONE,
         DLI-FUNCROLL.
```

Now let's modify the initialization logic to handle either a normal start or a restart. On a normal start we'll get the first root in the database. On a restart we'll reposition at the EMP-ID saved in the checkpoint that we are using to do the restart.

```
        CALL 'CBLTDLI' USING SIX,
             DLI-FUNCXRST,
             PCB-MASK,
             IO-AREALEN,
             XRST-IOAREA,
```

174

```
                    IMS-CHKP-AREA-LTH,
                    IMS-CHKP-AREA

           IF STATUS-CODE NOT EQUAL SPACES THEN
               PERFORM P9000-DISPLAY-ERROR
               GOBACK
           END-IF

           IF XRST-ID NOT EQUAL SPACES THEN
               SET SW-IMS-RESTART TO TRUE
               MOVE CHKP-NBR-LAST TO CHKP-NBR
               DISPLAY '*** COBIMSE IMS RESTART ***'
               DISPLAY '*   LAST CHECK POINT :' XRST-ID
               DISPLAY '*   EMPLOYEE NUMBER  :' CHKP-EMP-ID
           ELSE
               DISPLAY '****** COBIMSE IMS NORMAL START ***'
               PERFORM P8000-TAKE-CHECKPOINT
           END-IF.

       *   DO INITIAL DB READ FOR FIRST EMPLOYEE RECORD
       *   OR REPOSITION IF AN IMS RESTART.

           IF SW-IMS-RESTART THEN
               MOVE CHKP-EMP-ID TO EMP-ID-VAL
               PERFORM P2000-RESET-POSITION
           ELSE
               PERFORM P1000-GET-FIRST-ROOT

           END-IF.
```

The value of XRST-ID will be non-blank if we are doing a restart. In that case we will turn on the SW-IMS-RESTART switch. Otherwise we will branch to take the initial checkpoint. Now if the SW-IMS-RESTART is true, it means this is a restart so we load the employee id from the checkpoint area into the qualified EMPLOYEE qualified SSA value, and then we call the procedure to reset database position to where it was at that checkpoint.

If the value of XRST-ID is blank, then we are **not** doing a restart. In this case, we call the procedure to get the first root.

Finally, let's add a temporary statement to the execution loop. After a successful GHN, check to see if we have employee id 7459, and if so call the rollback procedure. We will only do this on the first run of the program so as to force a rollback.

```
DISPLAY 'SUCCESSFUL GET HOLD :'
   IO-EMPLOYEE-RECORD
MOVE EMP-ID TO EMP-ID-VAL
ADD +1 TO CHKP-COUNT
IF CHKP-COUNT GREATER THAN OR EQUAL TO 5
   PERFORM P8000-TAKE-CHECKPOINT
   PERFORM P2000-RESET-POSITION
END-IF
IF EMP-ID = '7459'
   PERFORM P3000-ROLLBACK
   GOBACK
END-IF
```

So here is our complete code listing. Review it carefully to be sure you understand what is happening.

```
IDENTIFICATION DIVISION.
PROGRAM-ID. COBIMSE.

*******************************************************
*  WALK THROUGH THE EMPLOYEE (ROOT) SEGMENTS OF       *
*  THE ENTIRE EMPLOYEE IMS DATABASE, AND ROLL BACK    *
*  CHNGES WHEN A PARTICULAR CONDITION IS ENCOUNTERED  *
*******************************************************

ENVIRONMENT DIVISION.
INPUT-OUTPUT SECTION.
DATA DIVISION.

*******************************************************
*  W O R K I N G   S T O R A G E   S E C T I O N   *
*******************************************************

WORKING-STORAGE SECTION.

 01 WS-FLAGS.
    05  SW-END-OF-DB-SWITCH    PIC X(1) VALUE 'N'.
        88  SW-END-OF-DB                VALUE 'Y'.
        88  SW-NOT-END-OF-DB            VALUE 'N'.

    05  SW-IMS-RESTART-SW      PIC X(1) VALUE 'N'.
        88  SW-IMS-RESTART              VALUE 'Y'.
        88  SW-NOT-IMS-RESTART          VALUE 'N'.

 01 DLI-FUNCTIONS.
    05 DLI-FUNCISRT  PIC X(4) VALUE 'ISRT'.
    05 DLI-FUNCGU    PIC X(4) VALUE 'GU '.
```

176

```
05 DLI-FUNCGN    PIC X(4) VALUE 'GN  '.
05 DLI-FUNCGHU   PIC X(4) VALUE 'GHU '.
05 DLI-FUNCGHN   PIC X(4) VALUE 'GHN '.
05 DLI-FUNCGNP   PIC X(4) VALUE 'GNP '.
05 DLI-FUNCREPL  PIC X(4) VALUE 'REPL'.
05 DLI-FUNCDLET  PIC X(4) VALUE 'DLET'.
05 DLI-FUNCXRST  PIC X(4) VALUE 'XRST'.
05 DLI-FUNCCHKP  PIC X(4) VALUE 'CHKP'.
05 DLI-FUNCROLL  PIC X(4) VALUE 'ROLL'.


01 IO-EMPLOYEE-RECORD.
   05  EMP-ID        PIC X(04).
   05  FILLER        PIC X(01).
   05  EMPL-LNAME    PIC X(30).
   05  FILLER        PIC X(01).
   05  EMPL-FNAME    PIC X(20).
   05  FILLER        PIC X(01).
   05  EMPL-YRS-SRV  PIC X(02).
   05  FILLER        PIC X(01).
   05  EMPL-PRM-DTE  PIC X(10).
   05  FILLER        PIC X(10).


01 EMP-UNQUALIFIED-SSA.
   05  SEGNAME    PIC X(08) VALUE 'EMPLOYEE'.
   05  FILLER     PIC X(01) VALUE ' '.


01 EMP-QUALIFIED-SSA.
   05  SEGNAME    PIC X(08) VALUE 'EMPLOYEE'.
   05  FILLER     PIC X(01) VALUE '('.
   05  FIELD      PIC X(08) VALUE 'EMPID'.
   05  OPER       PIC X(02) VALUE ' ='.
   05  EMP-ID-VAL PIC X(04) VALUE '    '.
   05  FILLER     PIC X(01) VALUE ')'.

01 SEG-IO-AREA    PIC X(80).


01 IMS-RET-CODES.
   05 ONE       PIC S9(9) COMP VALUE +1.
   05 TWO       PIC S9(9) COMP VALUE +2.
   05 THREE     PIC S9(9) COMP VALUE +3.
   05 FOUR      PIC S9(9) COMP VALUE +4.
   05 FIVE      PIC S9(9) COMP VALUE +5.
   05 SIX       PIC S9(9) COMP VALUE +6.


01 XRST-IOAREA.
```

```
      05 XRST-ID        PIC X(08) VALUE SPACES.
      05 FILLER         PIC X(04) VALUE SPACES.


  77 IO-AREALEN         PIC S9(9) USAGE IS BINARY VALUE 12.

  77 CHKP-ID            PIC X(08) VALUE 'IMSE-   '.


  77 CHKP-NBR           PIC 999   VALUE ZERO.
  77 CHKP-COUNT         PIC S9(9) USAGE IS BINARY VALUE ZERO.


  01 CHKP-MESSAGE.
      05 FILLER                PIC X(24) VALUE
         'COBIMSE  CHECK POINT NO:'.
      05 CHKP-MESS-NBR         PIC 999     VALUE ZERO.
      05 FILLER                PIC X(15)   VALUE '     ,AT REC#:'.
      05 FILLER                PIC X(10)   VALUE ' ,AT EMP#:'.
      05 CHKP-MESS-EMP         PIC X(04)   VALUE SPACES.


  01 IMS-CHKP-AREA-LTH.
      05 LEN                PIC S9(9) USAGE IS BINARY VALUE +7.


  01 IMS-CHKP-AREA.
      05 CHKP-EMP-ID     PIC X(04) VALUE SPACES.
      05 CHKP-NBR-LAST   PIC 999   VALUE 0.


  LINKAGE SECTION.

  01 IO-PCB.
      05 FILLER          PICTURE X(10).
      05 IO-STATUS-CODE  PICTURE XX.
      05 FILLER          PICTURE X(20).

  01 PCB-MASK.
      03 DBD-NAME        PIC X(8).
      03 SEG-LEVEL       PIC XX.
      03 STATUS-CODE     PIC XX.
      03 PROC-OPT        PIC X(4).
      03 FILLER          PIC X(4).
      03 SEG-NAME        PIC X(8).
      03 KEY-FDBK        PIC S9(5) COMP.
      03 NUM-SENSEG      PIC S9(5) COMP.
      03 KEY-FDBK-AREA.
         05 EMPLOYEE-KEY  PIC X(04).
         05 EMPPAYHS-KEY  PIC X(08).
```

```
PROCEDURE DIVISION.

    INITIALIZE IO-PCB PCB-MASK
    ENTRY 'DLITCBL' USING IO-PCB, PCB-MASK

    PERFORM P100-INITIALIZATION.
    PERFORM P200-MAINLINE.
    PERFORM P300-TERMINATION.
    GOBACK.


P100-INITIALIZATION.

    DISPLAY '** PROGRAM COBIMSE START **'
    DISPLAY 'PROCESSING IN P100-INITIALIZATION'.

* CHECK FOR RESTART

    CALL 'CBLTDLI' USING SIX,
         DLI-FUNCXRST,
         PCB-MASK,
         IO-AREALEN,
         XRST-IOAREA,
         IMS-CHKP-AREA-LTH,
         IMS-CHKP-AREA

    IF STATUS-CODE NOT EQUAL SPACES THEN
       PERFORM P9000-DISPLAY-ERROR
       GOBACK
    END-IF


    IF XRST-ID NOT EQUAL SPACES THEN
       SET SW-IMS-RESTART TO TRUE
       MOVE CHKP-NBR-LAST TO CHKP-NBR
       DISPLAY '*** COBIMSE IMS RESTART ***'
       DISPLAY '*  LAST CHECK POINT :' XRST-ID
       DISPLAY '*  EMPLOYEE NUMBER  :' CHKP-EMP-ID
    ELSE
       DISPLAY '****** COBIMSE IMS NORMAL START ***'
       PERFORM P8000-TAKE-CHECKPOINT
    END-IF.

*    DO INITIAL DB READ FOR FIRST EMPLOYEE RECORD
*    OR REPOSITION IF AN IMS RESTART.

    IF SW-IMS-RESTART THEN
       MOVE CHKP-EMP-ID TO EMP-ID-VAL
```

```
            PERFORM P2000-RESET-POSITION
        ELSE
            PERFORM P1000-GET-FIRST-ROOT

        END-IF.


    P200-MAINLINE.

        DISPLAY 'PROCESSING IN P200-MAINLINE'

*       CHECK STATUS CODE AND FIRST RECORD

        IF SW-END-OF-DB THEN
            DISPLAY 'NO RECORDS TO PROCESS!!'
        ELSE

            PERFORM UNTIL SW-END-OF-DB

                CALL 'CBLTDLI' USING THREE,
                     DLI-FUNCDLET,
                     PCB-MASK,
                     IO-EMPLOYEE-RECORD

                IF STATUS-CODE NOT EQUAL SPACES THEN
                    PERFORM P9000-DISPLAY-ERROR
                    GOBACK
                ELSE
                    DISPLAY 'SUCCESSFUL DELETE OF EMPLOYEE ' EMP-ID
                    MOVE EMP-ID TO CHKP-EMP-ID
                END-IF

*       GET THE NEXT RECORD

                CALL 'CBLTDLI' USING FOUR,
                     DLI-FUNCGHN,
                     PCB-MASK,
                     IO-EMPLOYEE-RECORD,
                     EMP-UNQUALIFIED-SSA

                IF STATUS-CODE = 'GB' THEN
                    SET SW-END-OF-DB TO TRUE
                    DISPLAY 'END OF DATABASE'
                ELSE
                    IF STATUS-CODE NOT EQUAL SPACES THEN
                        PERFORM P9000-DISPLAY-ERROR
                        SET SW-END-OF-DB TO TRUE
                        GOBACK
                    ELSE
                        DISPLAY 'SUCCESSFUL GET HOLD :'
```

```
                    IO-EMPLOYEE-RECORD
                MOVE EMP-ID TO EMP-ID-VAL
                ADD +1 TO CHKP-COUNT
                IF CHKP-COUNT GREATER THAN OR EQUAL TO 5
                    PERFORM P8000-TAKE-CHECKPOINT
                    PERFORM P2000-RESET-POSITION
                END-IF
                IF EMP-ID = '7459'
                    PERFORM P3000-ROLLBACK
                END-IF
            END-IF
        END-IF

    END-PERFORM.

    DISPLAY 'FINISHED PROCESSING IN P200-MAINLINE'.


P300-TERMINATION.

    DISPLAY 'PROCESSING IN P300-TERMINATION'
    ADD +1 TO CHKP-COUNT
    PERFORM P8000-TAKE-CHECKPOINT
    DISPLAY '** COBIMSE - SUCCESSFULLY ENDED **'.


P1000-GET-FIRST-ROOT.

    CALL 'CBLTDLI' USING FOUR,
        DLI-FUNCGHN,
        PCB-MASK,
        IO-EMPLOYEE-RECORD,
        EMP-UNQUALIFIED-SSA

    IF STATUS-CODE = '  ' THEN
        NEXT SENTENCE
    ELSE
        IF STATUS-CODE = 'GB' THEN
            SET SW-END-OF-DB TO TRUE
            DISPLAY 'END OF DATABASE :'
        ELSE
            PERFORM P9000-DISPLAY-ERROR
            GOBACK
        END-IF.

P2000-RESET-POSITION.

    DISPLAY 'PROCESSING IN P2000-RESET-POSITION'

    CALL 'CBLTDLI' USING ,
```

```
                    DLI-FUNCGHU,
                    PCB-MASK,
                    IO-EMPLOYEE-RECORD,
                    EMP-QUALIFIED-SSA

        IF STATUS-CODE NOT EQUAL SPACES THEN
            PERFORM P9000-DISPLAY-ERROR
            GOBACK
        ELSE
            DISPLAY 'SUCCESSFUL REPOSITION AT EMP ID ' EMP-ID.


    P3000-ROLLBACK.

        DISPLAY 'PROCESSING IN P3000-ROLLBACK'.

        CALL 'CBLTDLI' USING ONE,
             DLI-FUNCROLL.


    P8000-TAKE-CHECKPOINT.

        DISPLAY 'PROCESSING IN P8000-TAKE-CHECKPOINT'
        ADD +1             TO CHKP-NBR
        MOVE CHKP-NBR      TO CHKP-NBR-LAST
        MOVE CHKP-NBR-LAST TO CHKP-ID(6:3)
        DISPLAY 'CHECKPOINT ID IS ' CHKP-ID
        MOVE EMP-ID        TO CHKP-EMP-ID

        CALL 'CBLTDLI' USING SIX,
             DLI-FUNCCHKP,
             IO-PCB,
             IO-AREALEN,
             CHKP-ID,
             IMS-CHKP-AREA-LTH,
             IMS-CHKP-AREA

        IF IO-STATUS-CODE NOT EQUAL SPACES THEN
            DISPLAY 'TOOK AN ERROR DOING THE CHECKPOINT'
            DISPLAY 'IO-STATUS-CODE ' IO-STATUS-CODE
            PERFORM P9000-DISPLAY-ERROR
            GOBACK
        ELSE
            MOVE 0 TO CHKP-COUNT
            MOVE CHKP-NBR          TO CHKP-MESS-NBR
            MOVE CHKP-EMP-ID       TO CHKP-MESS-EMP
            DISPLAY CHKP-MESSAGE
        END-IF.

    P9000-DISPLAY-ERROR.
```

```
            DISPLAY 'ERROR ENCOUNTERED - DETAIL FOLLOWS'
            DISPLAY 'SEG-IO-AREA      :' SEG-IO-AREA
            DISPLAY 'DBD-NAME1:'      DBD-NAME
            DISPLAY 'SEG-LEVEL1:'     SEG-LEVEL
            DISPLAY 'STATUS-CODE:'    STATUS-CODE
            DISPLAY 'PROC-OPT1 :'     PROC-OPT
            DISPLAY 'SEG-NAME1 :'     SEG-NAME
            DISPLAY 'KEY-FDBK1 :'     KEY-FDBK
            DISPLAY 'NUM-SENSEG1:'    NUM-SENSEG
            DISPLAY 'KEY-FDBK-AREA1:' KEY-FDBK-AREA.

    *    END OF SOURCE CODE
```

Compile and link, then run the program. The program will abend with IMS user code **U0778** because of the ROLL call.[7] Here is the output:

```
** PROGRAM COBIMSE START **
PROCESSING IN P100-INITIALIZATION
****** COBIMSE IMS NORMAL START ***
PROCESSING IN P8000-TAKE-CHECKPOINT
CHECKPOINT ID IS IMSE-001
COBIMSE  CHECK POINT NO:001    ,AT REC#:  ,AT EMP#:
PROCESSING IN P200-MAINLINE
SUCCESSFUL DELETE OF EMPLOYEE 1111
SUCCESSFUL GET HOLD :1122 JENKINS                    DEBORAH          05
SUCCESSFUL DELETE OF EMPLOYEE 1122
SUCCESSFUL GET HOLD :3217 JOHNSON                    EDWARD           04
SUCCESSFUL DELETE OF EMPLOYEE 3217
SUCCESSFUL GET HOLD :4175 TURNBULL                   FRED             01
SUCCESSFUL DELETE OF EMPLOYEE 4175
SUCCESSFUL GET HOLD :4720 SCHULTZ                    TIM              09
SUCCESSFUL DELETE OF EMPLOYEE 4720
SUCCESSFUL GET HOLD :4836 SMITH                      SANDRA           03
PROCESSING IN P8000-TAKE-CHECKPOINT
CHECKPOINT ID IS IMSE-002
COBIMSE  CHECK POINT NO:002    ,AT REC#:  ,AT EMP#:4836
PROCESSING IN P2000-RESET-POSITION
SUCCESSFUL REPOSITION AT EMP ID 4836
SUCCESSFUL DELETE OF EMPLOYEE 4836
SUCCESSFUL GET HOLD :6288 WILLARD                    JOE              06
SUCCESSFUL DELETE OF EMPLOYEE 6288
SUCCESSFUL GET HOLD :7459 STEWART                    BETTY            07
PROCESSING IN P3000-ROLLBACK
```

---

7 If you prefer not to take a hard abend, instead of issuing the ROLL IMS call you can issue ROLB. ROLB backs out changes the same as ROLL, but ROLB returns control to the application program instead of abending.

At this point, we can verify that the first 5 records got deleted, and we can also verify that after the last checkpoint, all deleted records were backed (meaning they are still on the database).

```
Browse              USER01.IMS.EMPLOYEE.DATA                    Top of 4
Command ===>                                                    Scroll PAGE
                         Type DATA      RBA                     Format CHAR
                                            Col 1
----+----10---+----2----+----3----+----4----+----5----+----6----+----7----+----
****  Top of data  ****
......4836 SMITH                      SANDRA              03 2017-01-01 02
......6288 WILLARD                    JOE                 06 2016-01-01 20
......7459 STEWART                    BETTY               07 2016-07-31 01
......9134 FRANKLIN                   BRIANNA             00 2016-10-01 93
****  End of data  ****
```

The next step is to remove the code in COBIMSE that forced the rollback, then restart the program. Go ahead and remove or comment out the code, recompile and then we'll set up our restart JCL.

The PARM should look like this. Note that the IMSE-002 is the last successful checkpoint in the prior run. You can verify this by looking at the output from the previous run. Here is our restart parm override:

```
//GO       EXEC PGM=DFSRRC00,REGION=4M,
//         PARM=(DLI,&MBR,&PSB,7,0000,,0,'IMSE-002',N,0,0,,,N,N,,N,)
```

Now run the program, and here is the output.

```
** PROGRAM COBIMSE START **
PROCESSING IN P100-INITIALIZATION
*** COBIMSE IMS RESTART ***
*   LAST CHECK POINT :IMSE-002
*   EMPLOYEE NUMBER   :4836
PROCESSING IN P2000-RESET-POSITION
SUCCESSFUL REPOSITION AT EMP ID 4836
PROCESSING IN P200-MAINLINE
SUCCESSFUL DELETE OF EMPLOYEE 4836
SUCCESSFUL GET HOLD :6288 WILLARD                  JOE              06
SUCCESSFUL DELETE OF EMPLOYEE 6288
SUCCESSFUL GET HOLD :7459 STEWART                  BETTY            07
SUCCESSFUL DELETE OF EMPLOYEE 7459
SUCCESSFUL GET HOLD :9134 FRANKLIN                 BRIANNA          00
SUCCESSFUL DELETE OF EMPLOYEE 9134
END OF DATABASE
```

```
FINISHED PROCESSING IN P200-MAINLINE
PROCESSING IN P300-TERMINATION
PROCESSING IN P8000-TAKE-CHECKPOINT
CHECKPOINT ID IS IMSE-003
COBIMSE  CHECK POINT NO:003     ,AT REC#:  ,AT EMP#:9134
** COBIMSE - SUCCESSFULLY ENDED **
```

We correctly restarted at employee id 4836, and then processed in GHN mode from
there on. This is what should have happened. Now the database is empty, which we
can confirm by trying to browse it.

```
VSAM POINT RC X"08", Error Code X"20"
VSAM GET RC X"08", Error Code X"58"
Function terminated
***
```

You now have a basic model for doing checkpoint restart. Frankly, checkpoint restart
is done somewhat differently in each of the major environments I've worked in. Typically larger companies use third party products (such as BMC tools) to keep track of
checkpoints and facilitate recovery. You may need to learn a bit more to use the third
party products. The examples I've provided, although plain vanilla, work fine without
any third party products.

Here's the PLI version of this program.

```
PLIIMSE: PROCEDURE (IO_PTR_PCB,DB_PTR_PCB) OPTIONS(MAIN);
/*****************************************************************
* PROGRAM NAME: PLIIMSE - WALK THROUGH THE EMPLOYEE (ROOT)       *
*                         SEGMENTS OF THE EMPLOYEE IMS DATABASE,  *
*                         AND DELETE EACH ONE. ROLL BACK CHANGES  *
*                         WHEN A PARTICULAR EMP ID IS ENCOUNTERED. *
*****************************************************************/
/*****************************************************************
/*                W O R K I N G   S T O R A G E                  *
*****************************************************************/
    DCL SW_END_OF_DB             STATIC BIT(01) INIT('0'B);
    DCL SW_IMS_RESTART           STATIC BIT(01) INIT('0'B);
    DCL ONCODE                   BUILTIN;
    DCL DB_PTR_PCB               POINTER;
    DCL IO_PTR_PCB               POINTER;

    DCL PLITDLI                  EXTERNAL ENTRY;

    DCL 01 DLI_FUNCTIONS,
```

```
      05 DLI_FUNCISRT         CHAR(04) INIT ('ISRT'),
      05 DLI_FUNCGU           CHAR(04) INIT ('GU  '),
      05 DLI_FUNCGN           CHAR(04) INIT ('GN  '),
      05 DLI_FUNCGHU          CHAR(04) INIT ('GHU '),
      05 DLI_FUNCGNP          CHAR(04) INIT ('GNP '),
      05 DLI_FUNCREPL         CHAR(04) INIT ('REPL'),
      05 DLI_FUNCDLET         CHAR(04) INIT ('DLET'),
      05 DLI_FUNCXRST         CHAR(04) INIT ('XRST'),
      05 DLI_FUNCCHKP         CHAR(04) INIT ('CHKP'),
      05 DLI_FUNCROLL         CHAR(04) INIT ('ROLL');

DCL 01 IO_EMPLOYEE_RECORD,
      05  EMPL_ID            CHAR(04),
      05  FILLER1            CHAR(01),
      05  EMPL_LNAME         CHAR(30),
      05  FILLER2            CHAR(01),
      05  EMPL_FNAME         CHAR(20),
      05  FILLER3            CHAR(01),
      05  EMPL_YRS_SRV       CHAR(02),
      05  FILLER4            CHAR(01),
      05  EMPL_PRM_DTE       CHAR(10),
      05  FILLER5            CHAR(10);

DCL 01 EMP_UNQUALIFIED_SSA,
      05  SEGNAME            CHAR(08) INIT ('EMPLOYEE'),
      05  FILLER7            CHAR(01) INIT (' ');

DCL 01 EMP_QUALIFIED_SSA,
      05  SEGNAME            CHAR(08) INIT('EMPLOYEE'),
      05  FILLER8            CHAR(01) INIT('('),
      05  FIELD              CHAR(08) INIT('EMPID'),
      05  OPER               CHAR(02) INIT(' ='),
      05  EMP_ID_VAL         CHAR(04) INIT('    '),
      05  FILLER9            CHAR(01) INIT(')');

DCL ONE                      FIXED BIN (31) INIT(1);
DCL TWO                      FIXED BIN (31) INIT(2);
DCL THREE                    FIXED BIN (31) INIT(3);
DCL FOUR                     FIXED BIN (31) INIT(4);
DCL FIVE                     FIXED BIN (31) INIT(5);
DCL SIX                      FIXED BIN (31) INIT(6);

DCL 01 XRST_IOAREA,
      05 XRST_ID    CHAR(08) INIT('        '),
      05 FILLER10   CHAR(04) INIT('    ');

DCL XRST_IO_AREALEN    FIXED BIN(31) INIT (12);
DCL IO_AREALEN         FIXED BIN(31) INIT (08);
```

```
DCL CHKP_ID              CHAR(08) INIT( 'IMSE-   ');
DCL CHKP_NBR             PIC '999' INIT('000');
DCL CHKP_COUNT           FIXED BIN (31) INIT(0);

DCL 01 CHKP_MESSAGE,
       05 FILLER11           CHAR(24)    INIT(
           'COBIMSE   CHECK POINT NO:'),
       05 CHKP_MESS_NBR      PIC '999',
       05 FILLER12           CHAR(15)    INIT( ',AT INPUT REC#:'),
       05 CHKP_MESS_REC      PIC 'ZZZZZ9',
       05 FILLER13           CHAR(10)    INIT(',AT EMP#:'),
       05 CHKP_MESS_EMP      CHAR(08)    INIT(' ');

DCL IMS_CHKP_AREA_LTH    FIXED BIN (31) INIT(07);

DCL 01 IMS_CHKP_AREA,
       05 CHKP_EMP_ID        CHAR(04)    INIT('0000'),
       05 CHKP_NBR_LAST      CHAR(03)    INIT('000');

DCL 01 IO_PCB               BASED(IO_PTR_PCB),
       05 FILLER97           CHAR(10)  INIT(' '),
       05 IO_STATUS_CODE     CHAR(02)  INIT (' ');

DCL 01 PCB_MASK             BASED(DB_PTR_PCB),
       05 DBD_NAME           CHAR(08),
       05 SEG_LEVEL          CHAR(02),
       05 STATUS_CODE        CHAR(02),
       05 PROC_OPT           CHAR(04),
       05 FILLER99           FIXED BIN (31),
       05 SEG_NAME           CHAR(08),
       05 KEY_FDBK           FIXED BIN (31),
       05 NUM_SENSEG         FIXED BIN (31),
       05 KEY_FDBK_AREA,
       10 KFB_EMPLOYEE_ID CHAR(04);

/*********************************************************************
/*            P R O G R A M   M A I N L I N E                 *
/*********************************************************************/

CALL P100_INITIALIZATION;
CALL P200_MAINLINE;
CALL P300_TERMINATION;

P100_INITIALIZATION: PROC;

   PUT SKIP LIST ('PLIIMSE: TRAVERSE EMPLOYEE DATABASE ROOT SEGS');
   PUT SKIP LIST ('PROCESSING IN P100_INITIALIZATION');
   IO_PCB   = ' ';
```

```
        PCB_MASK = '';
        IO_EMPLOYEE_RECORD  = '';

        /* CHECK FOR RESTART */

        CALL PLITDLI (SIX,
                      DLI_FUNCXRST,
                      IO_PCB,
                      XRST_IO_AREALEN,
                      XRST_IOAREA,
                      IMS_CHKP_AREA_LTH,
                      IMS_CHKP_AREA);

        IF IO_STATUS_CODE ¬= ' ' THEN
           DO;
               CALL P9000_DISPLAY_ERROR;
               RETURN;
           END;

        IF XRST_ID ¬= ' ' THEN
           DO;
               SW_IMS_RESTART = '1'B;
               CHKP_NBR = CHKP_NBR_LAST;
               PUT SKIP LIST ('*** COBIMSE IMS RESTART ***');
               PUT SKIP LIST ('*  LAST CHECK POINT :' || XRST_ID);
               PUT SKIP LIST ('*  EMPLOYEE NUMBER  :' || CHKP_EMP_ID);
           END;
        ELSE
           DO;
               PUT SKIP LIST ('****** COBIMSE IMS NORMAL START ***');
               CALL P8000_TAKE_CHECKPOINT;
           END;

        IF SW_IMS_RESTART THEN
           DO;
               EMP_ID_VAL = CHKP_EMP_ID;
               CALL P2000_RESET_POSITION;
           END;
        ELSE
           CALL P1000_GET_FIRST_ROOT;

END P100_INITIALIZATION;

P200_MAINLINE: PROC;

    /*  MAIN LOOP - CYCLE THROUGH ALL ROOT SEGMENTS IN THE DB,
                    DISPLAYING THE DATA RETRIEVED               */
```

```
IF SW_END_OF_DB THEN
   PUT SKIP LIST ('NO RECORDS TO PROCESS!!');
ELSE
   DO WHILE (¬SW_END_OF_DB);
      PUT SKIP LIST ('SUCCESSFUL GET-HOLD :'
         || EMPL_ID);

      /* DELETE THE SWGMENT */

      CALL PLITDLI (THREE,
                    DLI_FUNCDLET,
                    PCB_MASK,
                    IO_EMPLOYEE_RECORD);

      IF STATUS_CODE ¬= '  ' THEN
         DO;
            CALL P9000_DISPLAY_ERROR;
            RETURN;
         END;
      ELSE
         PUT SKIP LIST ('SUCCESSFUL DELETE OF EMP ' || EMPL_ID);

      /* NOW GET THE NEXT ROOT TO DELETE */

      CALL PLITDLI (FOUR,
                    DLI_FUNCGN,
                    PCB_MASK,
                    IO_EMPLOYEE_RECORD,
                    EMP_UNQUALIFIED_SSA);

      IF STATUS_CODE = '  ' THEN
         DO;
            PUT SKIP LIST ('SUCCESSFUL GET HOLD: ' || EMPL_ID);
            EMP_ID_VAL = EMPL_ID;
            CHKP_COUNT = CHKP_COUNT + 1;
            IF CHKP_COUNT >= 5 THEN
               DO;
                  CALL P8000_TAKE_CHECKPOINT;
                  CALL P2000_RESET_POSITION;
               END;
         /* IF EMPL_ID = '7459' THEN
               CALL P3000_ROLLBACK;   */
         END;
      ELSE
         IF STATUS_CODE = 'GB' THEN
            DO;
               SW_END_OF_DB = '1'B;
               PUT SKIP LIST ('** END OF DATABASE');
```

```
                     END;
               ELSE
                  DO;
                     CALL P9000_DISPLAY_ERROR;
                     RETURN;
                  END;

      END;  /* DO WHILE */

   PUT SKIP LIST ('FINISHED PROCESSING IN P200_MAINLINE');

END P200_MAINLINE;

P300_TERMINATION: PROC;

   PUT SKIP LIST ('PROCESSING IN P300_TERMINATION');

   CHKP_COUNT = CHKP_COUNT + 1;
   CALL P8000_TAKE_CHECKPOINT;

   PUT SKIP LIST ('PLIIMSE - SUCCESSFULLY ENDED');

END P300_TERMINATION;

P1000_GET_FIRST_ROOT: PROC;

   PUT SKIP LIST ('PROCESSING IN P1000_GET_FIRST_ROOT');

 /* DO INITIAL DB READ FOR FIRST EMPLOYEE RECORD */

   CALL PLITDLI (FOUR,
                 DLI_FUNCGHN,
                 PCB_MASK,
                 IO_EMPLOYEE_RECORD,
                 EMP_UNQUALIFIED_SSA);

   IF STATUS_CODE = '  ' THEN;
   ELSE
      IF STATUS_CODE = 'GB' THEN
         DO;
            SW_END_OF_DB = '1'B;
            PUT SKIP LIST ('** END OF DATABASE');
         END;
      ELSE
         DO;
            CALL P9000_DISPLAY_ERROR;
            RETURN;
         END;
```

```
END P1000_GET_FIRST_ROOT;

P2000_RESET_POSITION: PROC;

    PUT SKIP LIST ('PROCESSING IN P2000_RESET_POSITION');

    CALL PLITDLI (FOUR,
                  DLI_FUNCGHU,
                  PCB_MASK,
                  IO_EMPLOYEE_RECORD,
                  EMP_QUALIFIED_SSA);

    IF STATUS_CODE ¬= '  ' THEN
       DO;
          CALL P9000_DISPLAY_ERROR;
          RETURN;
       END;
    ELSE
       PUT SKIP LIST ('SUCCESSFUL REPOSITION AT EMP ID ' || EMPL_ID);

END P2000_RESET_POSITION;

P3000_ROLLBACK: PROC;

    PUT SKIP LIST ('PROCESSING IN P3000_ROLLBACK');

    CALL PLITDLI (ONE,
          DLI_FUNCROLL);

    IF IO_STATUS_CODE ¬= '  ' THEN
       DO;
          CALL P9000_DISPLAY_ERROR;
          RETURN;
       END;
    ELSE
       PUT SKIP LIST ('SUCCESSFUL ROLLBACK TO CHKPID ' || CHKP_ID);

END P3000_ROLLBACK;

P8000_TAKE_CHECKPOINT: PROC;

    PUT SKIP LIST ('PROCESSING IN P8000_TAKE_CHECKPOINT');

    CHKP_NBR              = CHKP_NBR + 1;
    CHKP_NBR_LAST         = CHKP_NBR;
    SUBSTR(CHKP_ID,6,3)   = CHKP_NBR_LAST;
    CHKP_EMP_ID           = EMPL_ID;
```

191

```
       PUT SKIP LIST ('IO_AREALEN ' || IO_AREALEN);
       PUT SKIP LIST ('IMS_CHKP_AREA_LTH ' || IMS_CHKP_AREA_LTH);

       PUT SKIP LIST ('CHKP_ID = ' || CHKP_ID);
       PUT SKIP LIST (' ');

       CALL PLITDLI (SIX,
                     DLI_FUNCCHKP,
                     IO_PCB,
                     IO_AREALEN,
                     CHKP_ID,
                     IMS_CHKP_AREA_LTH,
                     IMS_CHKP_AREA);

       IF IO_STATUS_CODE ¬= ' ' THEN
          DO;
             PUT SKIP LIST ('TOOK AN ERROR DOING THE CHECKPOINT');
             PUT SKIP LIST ('IO_STATUS_CODE ' || IO_STATUS_CODE);
             CALL P9000_DISPLAY_ERROR;
             RETURN;
          END;
       ELSE
          DO;
             CHKP_COUNT = 0;
             CHKP_MESS_NBR = CHKP_NBR;
             CHKP_MESS_EMP = CHKP_EMP_ID;
             PUT SKIP LIST (CHKP_MESSAGE);
          END;

END P8000_TAKE_CHECKPOINT;

P9000_DISPLAY_ERROR: PROC;

    PUT SKIP LIST ('ERROR ENCOUNTERED - DETAIL FOLLOWS');
    PUT SKIP LIST ('SEG_IO_AREA     :' || SEG_IO_AREA);
    PUT SKIP LIST ('DBD_NAME1:' ||  DBD_NAME);
    PUT SKIP LIST ('SEG_LEVEL1:' || SEG_LEVEL);
    PUT SKIP LIST ('STATUS_CODE:' || STATUS_CODE);
    PUT SKIP LIST ('PROC_OPT1 :' || PROC_OPT);
    PUT SKIP LIST ('SEG_NAME1 :' || SEG_NAME);
    PUT SKIP LIST ('KEY_FDBK1 :' || KEY_FDBK);
    PUT SKIP LIST ('NUM_SENSEG1:' || NUM_SENSEG);
    PUT SKIP LIST ('KEY_FDBK_AREA1:' || KEY_FDBK_AREA);

END P9000_DISPLAY_ERROR;

END PLIIMSE;
```

That pretty well wraps up basic IMS programming. There are plenty of other features you can use, but that will depend on your work environment. Every shop and application is different. Good luck with it, and enjoy!

IMS Programming Guidelines

Consider the COBOL and PLI code examples in this text to be my own guidelines for coding IMS programs. There are more formal guidelines provided by IBM on their web site. [8]

---

8 https://www.ibm.com/support/knowledgecenter/SSEPH2_12.1.0/com.ibm.ims12.doc.apg/ims_programguidelines.htm

IMS Questions

1. What is the name of the interface program you call from a COBOL program to perform IMS operations?

2. Here are some IMS return codes and . Explain briefly what each of them means: blank, GE, GB, II

3. What is an SSA?

4. Briefly explain these entities: DBD, PSB, PCB?

5. What is the use of CMPAT parameter in PSB ?

6. In IMS, what is the difference between a key field and a search field?

7. What does PROCOPT mean in a PCB?

8. What are the four basic parameters of a DLI retrieval call?

9. What are Qualified SSA and Unqualified SSA?

10. Which PSB parameter in a PSBGEN specifies the language in which the application program is written?

11. What does SENSEG stand for and how is it used in a PCB?

12. What storage mechanism/format is used for IMS index databases?

13. What are the DL/I commands to add, change and remove a segment?

14. What return code will you receive from IMS if the DL/I call was successful?

15. If you want to retrieve the last occurrence of a child segment under its parent, what command code could you use?

16. When would you use a GU call?

17. When would you use a GHU call?

18. What is the difference between running an IMS program as DLI and BMP ?

19. When would you use a GNP call?

20. Which IMS call is used to restart an abended program?

21. How do you establish parentage on a segment occurrence?

22. What is a checkpoint?

23. How do you update the primary key of an IMS segment?

24. Do you need to use a qualified SSA with REPL/DLET calls?

25. What is a root segment?

26. What are command codes?

Answers to IMS Questions

1. What is the name of the interface program you call from a COBOL program to perform IMS operations?

    CBLTDLI is the normal interface program for a COBOL program to access IMS.

2. Here are some IMS return codes and . Explain briefly what each of them means: blank, GE, GB, II

    Blank – successful operation
    GE – segment not found
    GB – end of database
    II – duplicate key, insert failed

3. What is an SSA?

    Segment Search Argument – it is used to select segments by name and to specify search criteria for specific segments.

4. Briefly explain these entities: DBD, PSB, PCB?

    A Database Description (DBD) specifies characteristics of a database. The name, parent, and length of each segment type in the database.

    A Program Specification Block (PSB) is the program view of one or more IMS databases. The PSB includes one or more program communication blocks (PCB) for each IMS database that the program needs access to.

    A Program Communication Block (PCB) specifies the database to be accessed, the processing options such as read-only or various updating options, and the database segments that can be accessed.

5. What is the use of CMPAT parameter in PSB ?

   It is required if you are going to run your program in Batch Mode Processing (BMP), that is - in the online region. If you always run the program in DL/I mode, you do not need the CMPAT. If you are going to run BMP, you need the CMPAT=YES specified in the PSB.

6. In IMS, what is the difference between a key field and a search field?

   A key field is used to make the record unique and to order the database. A search field is a field that is needed to search the database on but does not have to be unique and does not order the database. For example, an EMPLOYEE database might be keyed on unique EMP-NUMBER. A search field might be needed on PHONE-NUMBER or ZIP-CODE. Even though the database is not ordered by these fields, they can be made search fields to query the database.

7. What does PROCOPT mean in a PCB?

   The PROCOPT parameter specifies processing options that are allowed for this PCB when operating on a segment.

   The different PROCOPTs and their meaning are:

   G - Get segment from DB
   I - Insert segment into DB
   R - Replace segment
   D - Delete segment
   A - All the above operations

8. What are the four basic parameters of a DLI retrieval call?

- **Function**
- **PCB mask**
- **SSAs**
- **IO Area**

9. What are Qualified SSA and Unqualified SSA?

A qualified SSA specifies the segment type and the specific instance (key) of the segment to be returned. An unqualified SSA simply supplies the name of the segment type that you want to operate upon. You could use the latter if you don't care which specific segment you retrieve.

10. Which PSB parameter in a PSBGEN specifies the language in which the application program is written?

The LANG parameter specifies the language in which the application program is written. Examples:

```
LANG=COBOL
LANG=PLI
LANG=ASSEM
```

11. What does SENSEG stand for and how is it used in a PCB?

SENSEG is known as Segment Level Sensitivity. It defines the program's access to parts of the database and it is identified at the segment level. For example, PROCOPT=G on a SENSEG means the segment is read-only by this PCB.

12. What storage mechanism/format is used for IMS index databases?

IMS index databases must use VSAM KSDS.

13. What are the DL/I commands to add, change and remove a segment?

The following are the DL/I commands for adding, changing and removing a segment:

- `ISRT`
- `REPL`
- `DLET`

14. What return code will you receive from IMS if the DL/I call was successful?

IMS returns blanks/spaces in the PCB STATUS-CODE field when the call was successful.

15. If you want to retrieve the last occurrence of a child segment under its parent, what command code could you use?

Use the L command code to retrieve the last child segment under its parent. Incidentally, IMS ignores the L command code at the root level.

16. When would you use a GU call?

GU is used to retrieve a segment occurrence based on SSA supplied arguments.

17. When would you use a GHU call?

GHU (Get Hold Unique) retrieves and locks the record that you intend to update or delete.

18. What is the difference between running an IMS program as DLI and BMP ?

DLI runs within its own address space. BMP runs under the IMS online control region. The practical difference concerns programs that update the database. If performing updates, DLI requires exclusive use of the database. Running BMP does not require exclusive use because it runs under control of the online region.

19. When would you use a GNP call?

The GNP call is used for Get Next within Parent. This function is used to retrieve segment occurrences in sequence subordinate to an established parent segment.

20. Which IMS call is used to restart an abended program?

The XRST IMS call is made to restart an abended IMS program. Assuming the program has taken checkpoints during the abended program execution, the XRST call is used to restart from the last checkpoint taken instead of starting the processing all over.

21. How do you establish parentage on a segment occurrence?

By issuing a successful GU or GN (or GHU or GHN) call that retrieves the segment on which the parentage is to be established. IMS normally sets parentage at the lowest level segment retrieved in a call. If you want to establish parentage at a level other than the normal level, use the P command code.

22. What is a checkpoint?

A checkpoint is a stage where the modifications done to a database by an application program are considered complete and are committed to the

database with the CKPT IMS call.

23. How do you update the primary key of an IMS segment?

You cannot update the primary key of a segment. If the key on a record must be changed, you can DLET the existing segment and then ISRT a new segment with the new key.

24. Do you need to use a qualified SSA with REPL/DLET calls?

No, you don't need to include an SSA with REPL/DLET calls. This is because the target segment has already been retrieved and held by a get hold call (that is the only way you can update or delete a segment).

25. What is a root segment?

A segment that lies at the top of the hierarchy is called the root segment. It is the only segment through which all dependent segments are accessed.

26. What are command codes?

Command codes are used along with SSAs to perform additional operations.

Common command codes used are:

'P' - used to set parentage on a particular segment.
'D' - used for path calls, to retrieve the entire hierarchical path.

Other IMS Resources

IBM Knowledge Centre
https://www.ibm.com/support/knowledgecenter/?mhq=ims%20documentation&mhsrc=ibmsearch_a

IMS Developer Community

https://developer.ibm.com/zsystems/ims/?mhq=ims%20documentation&mhsrc=ibmsearch_d

Other Titles by Robert Wingate

DB2 11 for z/OS Developer Training and Reference Guide

ISBN13: 9781734584707

This book will help you learn the basic and intermediate skills you need to write applications with DB2 11 for z/OS. The instruction, examples and questions/answers in this book are a fast track to becoming productive as quickly as possible. The content is easy to read and digest, well organized and focused on honing real job skills. DB2 11 for z/OS Developer Training and Reference Guide is a key step in the direction of mastering DB2 application development so you'll be ready to perform on a technical DB2 development team.

Db2 11 for LUW Developer Training and Reference Guide

ISBN13: 9781734584714
This book will help you learn the basic and intermediate skills you need to write applications with Db2 11 for Linux, UNIX and Windows. The instruction, examples and questions/answers in this book are a fast track to becoming productive as quickly as possible. The content is easy to read and digest, well organized and focused on honing real job skills. Demonstration programs are given in both the Java and c# .NET languages. Db2 11 for LUW Developer Training and Reference Guide is a key step in the direction of mastering Db2 application development so you'll be ready to perform on a technical development team.

IBM Mainframe Developer

Training and Reference Guide

JCL, MVS Utilities, COBOL, VSAM, IMS, DB2, CICS

Robert Wingate

ISBN: 9781734584738

This book will teach you the basic information and skills you need to develop applications on IBM mainframe computers running z/OS. The instruction, examples and sample programs in this book are a fast track to becoming productive as a developer in the IBM mainframe environment as quickly as possible. The coverage includes JCL, MVS Utilities, COBOL, VSAM, IMS, DB2 and CICS.

The content of this book is easy to read and digest, well organized and focused on honing real job skills. Acquiring these skills is a key step in mastering IBM application development so you'll be ready to perform effectively on an application development team.

CICS Basic Training for Application Developers Using DB2 and VSAM

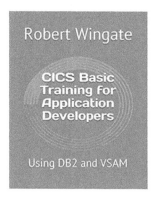

ISBN-13: 978-1794325067
This book will teach you the basic information and skills you need to develop applications with CICS on IBM mainframe computers running z/OS. The instruction, examples and sample programs in this book are a fast track to becoming productive as quickly as possible using CICS with the COBOL programming language. The content is easy to read and digest, well organized and focused on honing real job skills.

Quick Start Training for IBM z/OS Application Developers, Volume 1

ISBN-13: 978-1986039840
This book will teach you the basic information and skills you need to develop applications on IBM mainframes running z/OS. The instruction, examples and sample programs in this book are a fast track to becoming productive as quickly as possible in JCL, MVS Utilities, COBOL, PLI and DB2. The content is easy to read and digest, well organized and focused on honing real job skills. IBM z/OS Quick Start Training for Application Developers is a key step in the direction of mastering IBM application development so you'll be ready to join a technical team.

Quick Start Training for IBM z/OS Application Developers, Volume 2

ISBN-13: 978-1717284594
This book will teach you the basic information and skills you need to develop applications on IBM mainframes running z/OS. The instruction, examples and sample programs in this book are a fast track to becoming productive as quickly as possible in VSAM, IMS and DB2. The content is easy to read and digest, well organized and focused on honing real job skills. IBM z/OS Quick Start Training for Application Developers is a key step in the direction of mastering IBM application development so you'll be ready to join a technical team.

Teradata Basic Training for Application Developers

ISBN-13: 978-1082748882
This book will help you learn the basic information and skills you need to develop applications with Teradata. The instruction, examples and questions/answers in this book are a fast track to becoming productive as quickly as possible. The content is easy to read and digest, well organized and focused on honing real job skills. Programming examples are coded in both Java and C# .NET. Teradata Basic Training for Application Developers is a key step in the direction of mastering Teradata application development so you'll be ready to join a technical team.

DB2 Exam C2090-313 Preparation Guide

ISBN 13: 978-1548463052

This book will help you pass IBM Exam C2090-313 and become an IBM Certified Application Developer - DB2 11 for z/OS. The instruction, examples and questions/answers in the book offer you a significant advantage by helping you to gauge your readiness for the exam, to better understand the objectives being tested, and to get a broad exposure to the DB2 11 knowledge you'll be tested on.

DB2 Exam C2090-320 Preparation Guide

ISBN 13: 978-1544852096

This book will help you pass IBM Exam C2090-320 and become an IBM Certified Database Associate - DB2 11 Fundamentals for z/OS. The instruction, examples and questions/answers in the book offer you a significant advantage by helping you to gauge your readiness for the exam, to better understand the objectives being tested, and to get a broad exposure to the DB2 11 knowledge you'll be tested on. The book is also a fine introduction to DB2 for z/OS!

DB2 Exam C2090-313 Practice Questions

ISBN 13: 978-1534992467

This book will help you pass IBM Exam C2090-313 and become an IBM Certified Application Developer - DB2 11 for z/OS. The 180 questions and answers in the book (three full practice exams) offer you a significant advantage by helping you to gauge your readiness for the exam, to better understand the objectives being tested, and to get a broad exposure to the DB2 11 knowledge you'll be tested on.

DB2 Exam C2090-615 Practice Questions

ISBN 13: 978-1535028349

This book will help you pass IBM Exam C2090-615 and become an IBM Certified Database Associate (DB2 10.5 for Linux, Unix and Windows). The questions and answers in the book offer you a significant advantage by helping you to gauge your readiness for the exam, to better understand the objectives being tested, and to get a broad exposure to the knowledge you'll be tested on.

About the Author

Robert Wingate is a computer services professional with over 30 years of IBM mainframe programming experience. He holds several IBM certifications, including IBM Certified Application Developer - DB2 11 for z/OS, and IBM Certified Database Administrator for LUW. He lives in Fort Worth, Texas.

www.ingramcontent.com/pod-product-compliance
Lightning Source LLC
LaVergne TN
LVHW082033050326
832904LV00006B/272